Micromechanics of Defects in Solids

Pilar Ariza
Editor

Micromechanics of Defects in Solids

Previously published in *International Journal of Fracture*
Volume 174, Issue 1, 2012

Springer

Editor
Pilar Ariza
Universidad de Sevilla
Escuela Técnica Superior de Ingenieria
Camino de los Descubrimientos
s/n 41092 Sevilla
Spain

ISBN: 978-94-007-4625-1
Springer Dordrecht Heidelberg New York London

Library of Congress Control Number: 2012938216

Printed on acid-free paper

Springer is part of Springer Science+Business Media (www.springer.com)

Contents

Int J Fract (2012) 174:1
DOI 10.1007/s10704-012-9687-1

Preface

M. P. Ariza

Published online: 14 February 2012

Recent developments in experimental science that enable the examination of defects at the atomic scale provide an unprecedented connection between the structure and properties of materials. Techniques ranging from high-resolution electron microscopy to atomic-force microscopy reveal new insights into the micromechanical foundations of material behavior, but also pose deep challenges as regards theory, modeling and simulation. However, the link between the defects themselves and the observed macroscopic behavior is often a difficult one to forge theoretically or computationally and remains an active area of research. Many of the fundamental mechanisms underlying the inelastic behavior of materials are mediated by crystal-lattice defects and are, therefore, accessible to direct atomistic simulation, either by means of empirical potentials or through ab initio quantum-mechanical calculations. Notable examples are furnished by first-principles calculations of the EoS and elastic moduli of metals up to high pressures and temperatures, and the characterization of the structure of point defects, such as vacancies and interstitials, and extended defects, such as dislocations and grain boundaries. However, in general atomic-scale mechanisms are separated from macroscopic behavior by a vast array of intervening continuum scales. These mesoscopic scales both average and set the boundary conditions or driving forces for the atomic-scale phenomena and are an essential part of the structure of materials. While effective at describing macroscopic material behavior, continuum theories tend to break down on the scale of the lattice, e.g., in the vicinity of lattice defects. Therefore, a complete understanding of material behavior, as well as the predictive computation of the material properties, requires both atomistic and continuum modeling, with the atomistic/continuum handshake most effectively achieved within the framework of multiscale modeling.

This special issue constitutes a selection of the contributions presented at the International Symposium on Defect and Material Mechanics (ISDMM2011), held in Sevilla, Spain, June 27–July 1, 2011. All papers have been processed using according to the normal reviewing standards of the Journal. The ISDMM series of symposia provides a rare and much needed forum for bringing together a diverse group of researchers from various areas ranging from theoretical, experimental and computational modeling of the mechanics of materials. The present volume constitutes a valuable snapshot of the field of the mechanics of materials and their defects, and a window to its many accomplishments, challenges and opportunities, and open questions. It is especially to be desired that the volume will motivate the young research community interested in the field.

M. P. Ariza (✉)
Escuela Técnica Superior de Ingeniería,
Universidad de Sevilla, 41092 Seville, Spain
e-mail: mpariza@us.es

Int J Fract (2012) 174:3–15
DOI 10.1007/s10704-011-9664-0

Numerical simulation of intergranular and transgranular crack propagation in ferroelectric polycrystals

Amir Abdollahi · Irene Arias

Received: 15 September 2011 / Accepted: 29 November 2011 / Published online: 17 January 2012

Abstract We present a phase-field model to simulate intergranular and transgranular crack propagation in ferroelectric polycrystals. The proposed model couples three phase-fields describing (1) the polycrystalline structure, (2) the location of the cracks, and (3) the ferroelectric domain microstructure. Different polycrystalline microstructures are obtained from computer simulations of grain growth. Then, a phase-field model for fracture in ferroelectric single-crystals is extended to polycrystals by incorporating the differential fracture toughness of the bulk and the grain boundaries, and the different crystal orientations of the grains. Our simulation results show intergranular crack propagation in fine-grain microstructures, while transgranular crack propagation is observed in coarse grains. Crack deflection is shown as the main toughening mechanism in the fine-grain structure. Due to the ferroelectric domain switching mechanism, noticeable fracture toughness enhancement is also obtained for transgranular crack propagation. These observations agree with experiment.

Keywords Polycrystals · Ferroelectricity · Fracture · Phase-field models · Finite element analysis

A. Abdollahi · I. Arias (✉)
Laboratori de Càlcul Numèric (LaCàN), Departament de Matemàtica Aplicada III, Universitat Politècnica de Catalunya (UPC), Campus Nord UPC-C2, 08034 Barcelona, Spain
e-mail: irene.arias@upc.edu

1 Introduction

The design and implementation of electromechanical systems demand multifunctional materials with strong electromechanical coupling and reasonable reliability. In this category ferroelectric ceramics are the main candidates also exhibiting short response times. However, due to their inherent brittleness, a deep understanding of the fracture behavior of these materials is key to assure optimum reliability of the systems. Towards this goal, numerous theoretical and experimental investigations have been carried out during the past decades. Excellent reviews of these works are presented in (Zhang and Gao 2004; Schneider 2007; Kuna 2010). Ferroelectric ceramics exhibit ferroelectric and ferroelastic switching behavior with macroscopic dielectric and butterfly hystereses due to the underlying microscopic mechanisms. The reviews in (Kamlah 2001; Huber 2005) present related modeling approaches. The formation and evolution of the microstructure are more probable under high electromechanical loadings and near load concentrations such as the vicinity of cracks (Hackemann and Pfeiffer 2003; Jones et al. 2007). Interactions between the microstructure, grain boundaries, localized stress and electric fields near the crack tips lead to the complexity of fracture phenomena in ferroelectric polycrystals. Several approaches have been proposed to study the fracture in ferroelectric materials, including phenomenological constitutive models (Landis 2003; Sheng and Landis 2007; Wang and Landis 2006), models investigating the local phase transformations near the crack tip (Zhu and Yang 1997, 1999; Yang and Zhu 1998), phase-field models for computing the mechanical and electromechanical J-integrals (Song et al. 2007; Wang and Zhang 2007, 2008; Li and Landis 2011; Li and Kuna 2011), and more recently, a phase-field model accurately accounting for the stray fields (Yang and Dayal 2011). See also (DeSimone et al. 2001) for a related approach in ferromagnetics. However, these models assume fixed or stationary crack configurations and they are unable to capture the interactions between a propagating crack and the microstructure of the material. To tackle the full complexity of the phenomenon, we have recently introduced a phase-field model for the coupled microstructure and fracture evolution in ferroelectric single crystals (Abdollahi and Arias 2011b). We have shown that the interaction of the microstructure and the crack leads to a slow-fast crack propagation behavior observed in experiments. In another work, we have presented evidence that this interaction results in the anisotropy of the fracture toughness (Abdollahi and Arias 2011c). A related phase-field approach has been proposed for crack propagation and kinking in ferroelectrics, where the spontaneous rather than the total polarization of the material is chosen as the primary order parameter (Xu et al. 2010). In these phase-field models, both the crack and the domain walls are smeared and represented by internal layers. Therefore, the tracking of these interfaces and the modeling of the interfacial phenomena (domain wall energies, crack face boundary conditions) are automatically accomplished by the governing equations. Nevertheless, this flexibility comes at the expense of a high computational cost, since the width of the phase-field regularizations of the domain wall and the crack must be resolved by the discretization.

To the best of our knowledge, no model has been proposed in the literature to study the interactions between the fracture process and the polycrystalline and domain structure in ferroelectric polycrystals. The related models (Landis 2003; Li and Kuna 2011) have only taken into account the grains orientations, but the effects of grain boundaries and grain size have not been considered. In (Verhoosel and Gutierrez 2009), the grain microstructure is introduced to model inter- and transgranular crack propagation in linear piezoelectric polycrystals, thus not accounting for the effect of ferroelectric domain microstructures. The main objective of this paper is to study the fracture processes in ferroelectric polycrystals in its full complexity, and in particular to evaluate the effects produced by the grain and ferrolelectric domain microstructures on the fracture response of the material. For this purpose, we extend to polycrystals the phase-field model of fracture in ferroelectric single crystals that we have recently proposed (Abdollahi and Arias 2011a, b, c). The results obtained with this model prove its potential to capture complex interactions between the crack and the material microstructure. The theory of the model for polycrystals is summarized in Sect. 2. Numerical simulations are presented in Sect. 3, along with a qualitative discussion of the observed fracture mechanisms and their relation to the polycrystalline and the ferroelectric domain microstructures. The last section is the conclusion of this paper.

2 Theory

The proposed model for fracture of ferroelectric polycrystals is based on two existing phase-field approaches for: (1) grain growth (Fan and Chen 1997) and (2) fracture in ferroelectric single crystals (Abdollahi and Arias 2011b). The first model provides realistic polycrystalline microstructures with different grain sizes, while the second models the complex interactions between the fracture processes and the ferroelectric domain formation and evolution. Following (Zhang and Bhattacharya 2005b), the latter is extended to ferroelectric polycrystals. The resulting model couples three phase-fields describing (1) the polycrystalline structure, (2) the location of the cracks, and (3) the ferroelectric domain microstructure. In this section, the phase-field models for grain growth and fracture of ferroelectric single crystals are briefly discussed. Then, the extension to ferroelectric polycrystals and its numerical implementation are described.

2.1 Phase-field model for grain growth

According to the phase-field model of grain growth (Fan and Chen 1997), the total free energy of a heterogeneous system is stated as

$$F = \int_{\Omega} \left[f_0(\eta_1, \eta_2, \ldots, \eta_m) + \sum_{i=1}^{m} \frac{\kappa_i}{2} (\nabla \eta_i)^2 \right] \mathrm{d}\Omega, \tag{1}$$

where f_0 is the local free energy density associated with the orientation field variables $(\eta_1, \eta_2, \ldots, \eta_m)$ for distinguishing different orientations of grains and m is the number of possible orientations. κ_i are the coefficients of the gradient energy terms penalizing sharp variations in the field variables. The free energy density f_0 is written as (Fan and Chen 1997)

$$f_0(\eta_1, \eta_2, \ldots, \eta_m) = \sum_{i=1}^{m} \left(-\tfrac{\alpha}{2}\eta_i^2 + \tfrac{\beta}{4}\eta_i^4\right) + \gamma \sum_{i=1}^{m} \sum_{j\neq i}^{m} \eta_i^2 \eta_j^2, \tag{2}$$

where α, β and γ are positive constants characterizing the energy landscape. For $\gamma > \beta/2$, f_0 furnishes a multi-well energy landscape with $2m$ minima located at $(\eta_1, \eta_2, \ldots, \eta_m) = (1, 0,\ldots, 0), (0, 1,\ldots, 0),\ldots, (0, 0,\ldots, 1), (-1, 0,\ldots, 0), (0, -1, \ldots, 0), \ldots, (0, 0,\ldots, -1)$. Each of these $2m$ minima corresponds to a number of grains constructing a part of the polycrystalline microstructure. The grain boundaries are the regions of the domain where the gradient energy terms are non-zero. The evolution of the grains is governed by the time-dependent Ginzburg-Landau (TDGL) equations as

$$\mu_g \int_{\Omega} \dot{\eta}_i \delta\eta_i \mathrm{d}\Omega = -\delta F(\eta_i; \delta\eta_i) = \int_{\Omega} \left(\alpha\eta_i - \beta\eta_i^3 - 2\gamma\eta_i \sum_{j\neq i}^{m} \eta_i^2\right) \mathrm{d}\Omega + \int_{\Omega} (\kappa_i \nabla^2 \eta_i) \mathrm{d}\Omega, \tag{3}$$

where $i = 1, 2, \ldots, m$ and $1/\mu_g$ is the mobility of the process.

2.2 Phase-field model for fracture of ferroelectric single crystals

In the spirit of Francfort and Marigo (1998) and Shu and Bhattacharya (2001), the total electromechanical enthalpy of a possibly fractured ferroelectric single crystal occupying a region Ω is stated as (Abdollahi and Arias 2011b)

$$H[\mathbf{u}, v, \mathbf{p}, \phi] = \int_{\Omega} h(\boldsymbol{\varepsilon}(\mathbf{u}), \mathbf{p}, \nabla\mathbf{p}, \mathbf{E}(\phi), v)\, \mathrm{d}\Omega + G_c \int_{\Omega} \left[\frac{(1-v)^2}{4\kappa} + \kappa|\nabla v|^2\right] \mathrm{d}\Omega, \tag{4}$$

where body loads, volume charges, tractions and surface charges have been ignored for simplicity. The first integral is referred to as total bulk energy of the material while the second integral takes the role of the surface energy. The constant G_c is the critical energy release rate or the surface energy density in Griffith's theory (Griffith 1921). The scalar field v is the phase-field parameter describing a smooth transition in space between unbroken ($v = 1$) and broken ($v = 0$) states of the material. When the positive regularization parameter κ tends to zero, this transition becomes sharper. The electro-mechanical enthalpy density h is given as

$$h(\boldsymbol{\varepsilon}, \mathbf{p}, \nabla\mathbf{p}, \mathbf{E}, v) = (v^2 + \eta_\kappa)\left[U(\nabla\mathbf{p}) + W(\mathbf{p}, \boldsymbol{\varepsilon})\right] + \chi(\mathbf{p}) - \frac{\varepsilon_0}{2}|\mathbf{E}|^2 - \mathbf{E}\cdot\mathbf{p}, \tag{5}$$

where $\boldsymbol{\varepsilon}$ is the strain tensor associated with the mechanical displacement $\mathbf{u}$, $\boldsymbol{\varepsilon} = 1/2(\nabla\mathbf{u} + \nabla^T\mathbf{u})$, $\mathbf{p}$ is the polarization, $\mathbf{E}$ is the electric field associated with the electric potential ϕ as $\mathbf{E} = -\nabla\phi$ and ε_0 is the permittivity of free space. η_κ is a small (relative to κ) residual stiffness to avoid the singularity of the first part of the energy in fully fractured regions of the domain. U is the domain wall energy density penalizing sharp variations in the polarization, χ is the phase separation potential, and W is the electroelastic energy density. The energy densities χ and W penalize deviations from the spontaneous polarizations and strains of the material, hence introducing the anisotropy and nonlinearity of ferroelectric materials.

This particular formulation of the phase-field model encodes the following assumed crack conditions: (1) traction free, (2) electrically permeable, and (3) free polarization (Abdollahi and Arias 2011a). For a sharp crack, the traction-free conditions are stated as $\boldsymbol{\sigma}\cdot\mathbf{n} = \mathbf{0}$ where $\boldsymbol{\sigma}$ and $\mathbf{n}$ are the stress tensor and the unit outward normal on the crack faces, respectively. The permeable crack conditions assume that the crack does not perturb the electrical fields and the electric potential and the normal component of the electric displacement are continuous across the crack faces, i.e.

$$\phi^+ = \phi^- \quad \text{and} \quad \mathbf{D}^+\cdot\mathbf{n}^+ = \mathbf{D}^-\cdot\mathbf{n}^-, \tag{6}$$

where the superscripts + and − denote the top and bottom crack faces. The free-polarization boundary conditions imply that the gradients normal to the crack faces of the polarization components vanish at the crack faces, i.e.

$$\frac{dp_i^+}{dn} = \frac{dp_i^-}{dn} = 0, \ (i = 1, 2). \tag{7}$$

It can be shown numerically that all of these conditions are satisfied in the vicinity of the smeared crack for a finite but small value of the regularization parameter κ, as expected in the sharp crack model (Abdollahi and Arias 2011b). Other electro-mechanical crack conditions can be modeled similarly (Abdollahi and Arias 2011a).

The energy terms are chosen following (Devonshire 1949, 1951) for the case of plane polarization and plane strain as

$$U(p_{i,j}) = \frac{a_0}{2}(p_{1,1}^2 + p_{1,2}^2 + p_{2,1}^2 + p_{2,2}^2), \tag{8}$$

$$\begin{aligned} W(p_i, \varepsilon_{jk}) = & -\frac{b_1}{2}(\varepsilon_{11}p_1^2 + \varepsilon_{22}p_2^2) \\ & -\frac{b_2}{2}(\varepsilon_{11}p_2^2 + \varepsilon_{22}p_1^2) - b_3(\varepsilon_{21} + \varepsilon_{12}) \\ & \times p_1 p_2 + \frac{c_1}{2}(\varepsilon_{11}^2 + \varepsilon_{22}^2) + c_2\varepsilon_{11}\varepsilon_{22} \\ & + \frac{c_3}{2}(\varepsilon_{12}^2 + \varepsilon_{21}^2), \end{aligned} \tag{9}$$

$$\begin{aligned} \chi(p_i) = & \frac{a_1}{2}(p_1^2 + p_2^2) + \frac{a_2}{4}(p_1^4 + p_2^4) + \frac{a_3}{2}(p_1^2 p_2^2) \\ & + \frac{a_4}{6}(p_1^6 + p_2^6) + \frac{a_5}{4}(p_1^4 p_2^4), \end{aligned} \tag{10}$$

where a_0 is the scaling parameter of the domain wall energy, $a_i (i = 1, \ldots, 5)$ are the constants of the phase separation energy, $b_i (i = 1, 2, 3)$ are the constants of the electro-machanical coupling terms and $c_i (i = 1, 2, 3)$ are the elastic constants. The combination of the energy functions χ and W is the total Landau-Devonshire energy density furnishing a multi-well energy landscape with four minima corresponding to the four variants of the tetragonal phase. The time evolution of the governing equations along with a simple algorithm to solve the coupled system are presented in Sect. 2.3, for the polycrystal model.

2.3 Phase-field model for fracture of ferroelectric polycrystals

The total electromechanical enthalpy in Eq. (4) is extended here to study the crack propagation in ferroelectric polycrystals. For this purpose, the polycrystalline microstructure given by the model of grain growth described in Sect. 2.1 is included in the formulation as follows. In the models of brittle polycrystalline materials, the fracture properties of the grain boundaries can be characterized by assuming a lower fracture toughness along these boundaries as compared to the grain interiors. Other works (Grah et al. 1996; Sukumar et al. 2003; Verhoosel and Gutierrez 2009) have proven the validity of this assumption for the study of the intergranular and the transgranular modes of fracture. In the model of grain growth described in Sect. 2.1, the grain boundaries can be identified through a function ξ of the orientation field variables η_i defined as

$$\xi = \sum_{i=1}^{m} \eta_i^{2k}, \tag{11}$$

where k is a positive integer. The function ξ with $k = 1$ is commonly used to visualize the polycrystalline microstructures. It has a unit value inside the grains and smaller positive values at the grain boundaries (Fan and Chen 1997). Through this function, we can easily assign a lower fracture toughness along the grain boundaries than in the grain interiors. The critical energy release rate in a polycrystal can be defined as

$$G_c^{poly} = G_c\,\mathscr{F}(\xi), \tag{12}$$

where G_c is the critical energy release rate of the bulk crystal, and the function $\mathscr{F}$ controls the weakening of the material at the grain boundaries. This function attains its maximum for $\xi = 1$ (inside the grains) and its minimum indicates the ratio of the critical fracture energy of the grain boundary to that of the grain interior. By changing this ratio, it is possible to study the transition between the intergranular and transgranular modes of crack growth in polycrystalline materials (Sukumar et al. 2003). This ratio can be regulated in our model by an appropriate selection of the function $\mathscr{F}$ and setting the power k in Eq. (11) and the coefficients of the gradient energy (κ_i). Here, the function $\mathscr{F}$ is chosen to be proportional to ξ, i.e. $\mathscr{F}(\xi) = \xi$. The power of ξ is also assigned as $k = 1$ to avoid decreasing the fracture toughness towards zero along the grain boundaries, which is unphysical. Note that other modeling options are possible, but in the absence of precise information about the grain boundary properties, the choices of the function $\mathscr{F}$ and power k seem adequate.

We also assume that each grain is oriented along a different crystallographic direction. In 2D, the rotation angle Θ is defined and its value is assigned randomly for each orientation field variables $\eta_i (i = 1, \ldots, m)$. Then, the rotation field over a polycrystalline microstructure can be obtained as

$$\theta = \frac{\sum_{i=1}^{m} \Theta_i \eta_i^2}{\sum_{i=1}^{m} \eta_i^2}. \tag{13}$$

The function θ has the value Θ_i inside the grains corresponding to the orientation field variable η_i. At the grain boundaries, the numerator in Eq. (13) decreases significantly. The function ξ in the denominator compensates for this effect and assures a smooth transition between the orientations of adjacent grains across their boundaries.

Using the rotation field θ and following the phase-field model of ferroelectric polycrystals (Zhang and Bhattacharya 2005b), the electromechanical enthalpy density h in Eq. (5) can be extended to polycrystals. For this purpose, each of the energy terms in Eq. (5) is transformed from the local coordinate system of each individual grain to the global coordinate system of the polycrystal, θ being the angle between these two coordinate systems. It can be shown that the domain wall energy density U, the electroelastic energy density W and the last two terms in Eq. (5) remain unchanged by this transformation. The only modified energy term is the phase-separation potential χ which is converted to

$$
\begin{aligned}
\chi^{poly}(p_i) = {} & \frac{f_1}{2}(p_1^2+p_2^2)+\frac{f_2}{4}(p_1^4+p_2^4)+\frac{f_3}{2}p_1^2p_2^2 \\
& +\frac{f_4}{6}(p_1^6+p_2^6)+\frac{f_5}{4}p_1^4p_2^4 \\
& +g_1(p_1^3p_2-p_1p_2^3) \\
& +g_2(p_1^5p_2-p_1p_2^5)+\frac{g_3}{2}(p_1^4p_2^2+p_1^2p_2^4) \\
& +\frac{g_4}{8}(p_1^8+p_2^8)+g_5(p_1^7p_2-p_1p_2^7) \\
& +\frac{g_6}{2}(p_1^6p_2^2+p_1^2p_2^6)+g_7(p_1^5p_2^3-p_1^3p_2^5),
\end{aligned}
\tag{14}
$$

where

$$
\begin{aligned}
& f_1 = a_1, \quad f_2 = \frac{1}{4}[(3+\cos 4\theta)a_2+(1-\cos 4\theta)a_3], \\
& f_3 = \frac{1}{4}[3(1-\cos 4\theta)a_2+(1+3\cos 4\theta)a_3], \\
& f_4 = \frac{1}{8}(5+3\cos 4\theta)a_4, \\
& f_5 = \frac{1}{64}(9+20\cos 4\theta+35\cos 8\theta)a_5, \\
& g_1 = \frac{1}{4}(a_2-a_3)\sin 4\theta, \\
& g_2 = \frac{1}{4}a_4\sin 4\theta, \quad g_3 = \frac{5}{8}a_4(1-\cos 4\theta), \\
& g_4 = \frac{1}{64}a_5(3-4\cos 4\theta+\cos 8\theta), \\
& g_5 = \frac{1}{64}a_5(-2\sin 4\theta+\sin 8\theta), \\
& g_6 = \frac{1}{64}a_5(3+4\cos 4\theta-7\cos 8\theta), \\
& g_7 = \frac{1}{64}a_5(-2\sin 4\theta-7\sin 8\theta).
\end{aligned}
\tag{15}
$$

By replacing χ with χ^{poly} in Eq. (5), the electromechanical enthalpy density for a ferroelectric polycrystal takes the form

$$
\begin{aligned}
h^{poly}(\boldsymbol{\varepsilon},\mathbf{p},\nabla\mathbf{p},\mathbf{E},v) = {} & (v^2+\eta_\kappa)\left[U(\nabla\mathbf{p})+W(\mathbf{p},\boldsymbol{\varepsilon})\right] \\
& +\chi^{poly}(\mathbf{p})-\frac{\varepsilon_0}{2}|\mathbf{E}|^2-\mathbf{E}\cdot\mathbf{p}.
\end{aligned}
\tag{16}
$$

Finally, the total electromechanical enthalpy in Eq. (4) is formulated for a possibly fractured ferroelectric polycrystal by considering Eqs. (12) and (16) as

$$
\begin{aligned}
H[\mathbf{u},v,\mathbf{p},\phi] = {} & \int_\Omega h^{poly}(\boldsymbol{\varepsilon}(\mathbf{u}),\mathbf{p},\nabla\mathbf{p},\mathbf{E}(\phi),v)\,\mathrm{d}\Omega \\
& +G_c^{poly}\int_\Omega\left[\frac{(1-v)^2}{4\kappa}+\kappa|\nabla v|^2\right]\mathrm{d}\Omega.
\end{aligned}
\tag{17}
$$

The evolution of the ferroelectric domain microstructure and the fracture results from the gradient flows of the total electro-mechanical enthalpy based on the time-dependent Ginzburg-Landau (TDGL) equations. Here, the primary order variables are $\mathbf{p}$ and v for the microstructure and fracture processes, respectively. Thus, the governing equations are obtained by assuming that the displacement and the electric field adjust immediately to mechanical and electrostatic equilibrium (with infinite mobility), i.e.

$$\mu_p\int_\Omega \dot{p}_i\delta p_i\mathrm{d}\Omega = -\int_\Omega \frac{\partial h^{poly}}{\partial p_i}\delta p_i\,\mathrm{d}\Omega, \tag{18}$$

$$\mu_v\int_\Omega \dot{v}\delta v\mathrm{d}\Omega = -\int_\Omega \frac{\partial h^{poly}}{\partial v}\delta v\,\mathrm{d}\Omega, \tag{19}$$

$$0 = \int_\Omega \sigma_{ij}\delta\varepsilon_{ij}\,\mathrm{d}\Omega, \tag{20}$$

$$0 = \int_\Omega D_i\delta E_i\,\mathrm{d}\Omega, \tag{21}$$

where $1/\mu_p > 0$ and $1/\mu_v > 0$ are the mobilities of the processes. The stresses and electric displacements are derived from the enthalpy density, respectively, as $\boldsymbol{\sigma} = \partial h^{poly}/\partial\boldsymbol{\varepsilon}$ and $\mathbf{D} = -\partial h^{poly}/\partial\mathbf{E}$.

The weak forms of the evolution and equilibrium equations are discretized in space with standard finite elements. Equations (18) and (19) are discretized in time with a semi-implicit scheme. A simple algorithm to solve the coupled system in a straightforward staggered approach is presented next. This algorithm describes how to advance in one load step w^n, and it is meant to achieve steady states for both ferroelectric domains and fracture processes in each load step. The functions $\mathbf{g}(w^n)$ and $f(w^n)$ encode the Dirichlet data for the mechanical displacement and electric potential as a function of the load step. After reaching a steady state for both the polarization $\mathbf{p}$ and the phase-field v, the values for v^n, $\mathbf{u}^n$, $\mathbf{p}^n$ and ϕ^n are recorded and the algorithm proceeds to a new load step w^{n+1}.

Algorithm 1 for the coupled model

Let $m = 0$
Set $v_0 = 1$, $\mathbf{p}_0 = \mathbf{p}_{init}$, $\phi_0 = 0$ and $\mathbf{u}_0 = 0$ if $n = 0$
Set $v_0 = v^{n-1}$, $\mathbf{u}_0 = \mathbf{u}^{n-1}$, $\mathbf{p}_0 = \mathbf{p}^{n-1}$ and $\phi_0 = \phi^{n-1}$ if $n > 0$
repeat
  $m \longleftarrow m + 1$
  Compute $\mathbf{p}_m$ in Eq. (18) using $\mathbf{p}_{m-1}$, $\mathbf{u}_{m-1}$, ϕ_{m-1} and v_{m-1}
  Compute $\mathbf{u}_m$ in Eq. (20) using $\mathbf{p}_m$ and v_{m-1} under the constraint $\mathbf{u}_m = \mathbf{g}(w^n)$ on $\Gamma_{D,\mathbf{u}}$
  Compute ϕ_m in Eq. (21) using $\mathbf{p}_m$ under the constraint $\phi_m = f(w^n)$ on $\Gamma_{D,\phi}$
  Compute v_m in Eq. (19) using $\mathbf{p}_m$, $\mathbf{u}_m$ and v_{m-1} under the constraint $v_m = 0$ for $v^{n-1} \leqslant \zeta$
until $\|\mathbf{p}_m - \mathbf{p}_{m-1}\|_\infty \leqslant \delta_{ferro}$ and $\|v_m - v_{m-1}\|_\infty \leqslant \delta_{vfield}$
Set $\mathbf{u}^n = \mathbf{u}_m$, $v^n = v_m$, $\mathbf{p}^n = \mathbf{p}_m$ and $\phi^n = \phi_m$

3 Numerical simulations and discussion

We consider a rectangular domain of normalized dimensions 200×200 according to the normalizations presented in Table 1. It is discretized with approximately 80,000 triangular finite elements of size $h \simeq 1$. First, the phase-field model of grain growth described in Sect. 2.1 is used to obtain different polycrystalline microstructures. The parameters of the kinetic equation (3) are selected as $\alpha = 1$, $\beta = 1$, $\gamma = 1$, $\kappa_i = 0.5$ and $\mu_g = 1$. The time step $\Delta t = 0.25$ is also chosen for the numerical integration scheme. Twenty field variables ($m = 20$) are considered and they are initialized with random small variables as $-0.001 < \eta_i (i=1, \ldots, 20) < 0.001$. Periodic boundary conditions are applied. The simulation is carried out on parallel processors for 7,000 steps using the finite element library of the Kratos multi-physics package

Table 1 Normalization of the variables and parameters (i = 1,2,3)

x_i'	p_i'	t'	ε_0'	
$x_i\sqrt{c_0/a_0}/p_0$	p_i/p_0	$tc_0/\mu p_0^2$	$\varepsilon_0 c_0/p_0^2$	
ϕ'	a_1'	a_2'	a_3'	a_4'
$\phi/\sqrt{a_0 c_0}$	$a_1 p_0^2/c_0$	$a_2 p_0^4/c_0$	$a_3 p_0^4/c_0$	$a_4 p_0^6/c_0$
a_5'	b_i'	c_i'	G_c'	
$a_5 p_0^8/c_0$	$b_i p_0^2/c_0$	c_i/c_0	$G_c\sqrt{1/a_0 c_0}/p_0$	

(Dadvand et al. 2010). The contour of ξ in Eq. (11) is presented in Fig. 1 for two snapshots of the polycrystalline microstructure evolution. Grain boundaries are highlighted by darker regions in this figure. In the early stages of the simulation, the field variables η_i start to grow at different locations of the domain towards the crystallization. After about 500 steps, a well defined microstructure can be observed in Fig. 1a which consists of about 220 grains. Further evolution of the model leads to grain growth and coalescence. A coarse-grain microstructure is obtained at the end of the simulation which is depicted in Fig. 1b.

The two polycrystalline microstructures shown in Fig. 1a and b are selected for the fracture simulations in a fine- and a coarse-grain ferroelectric polycrystal, respectively. The function ξ has the lowest value about 0.2 at the grain boundaries for both micostructures. It has been shown that if this ratio is below 0.3, the fracture mode is entirely intergranular over a fine-grain microstructure (Sukumar et al. 2003). The grain orientations $\Theta_i (i = 1, \ldots, 20)$ are assigned randomly between 0° and 45° using the function θ in Eq. (13).

For the fracture simulations, a monotonically increasing mechanical load is applied by pulling the top and bottom sides of the model with a uniform vertical mechanical displacement such that $\mathbf{u}_\pm = (0, \pm w)$, where + and − indicate the top and bottom sides of the model respectively and w is the load step, see Fig. 2. The vertical mechanical displacement is also constrained at the top half and bottom half of the left side such that $u_{2\pm} = \pm w$. As this model does not have any pre-crack, this extra boundary condition is needed to force the crack to initiate at the center-left of the model. For all the simulations, the initial polarization $\mathbf{p}_{init} = (1, 0)$ is assigned along the positive x_1 direction in the global coordinate system, see Fig. 2a. Note

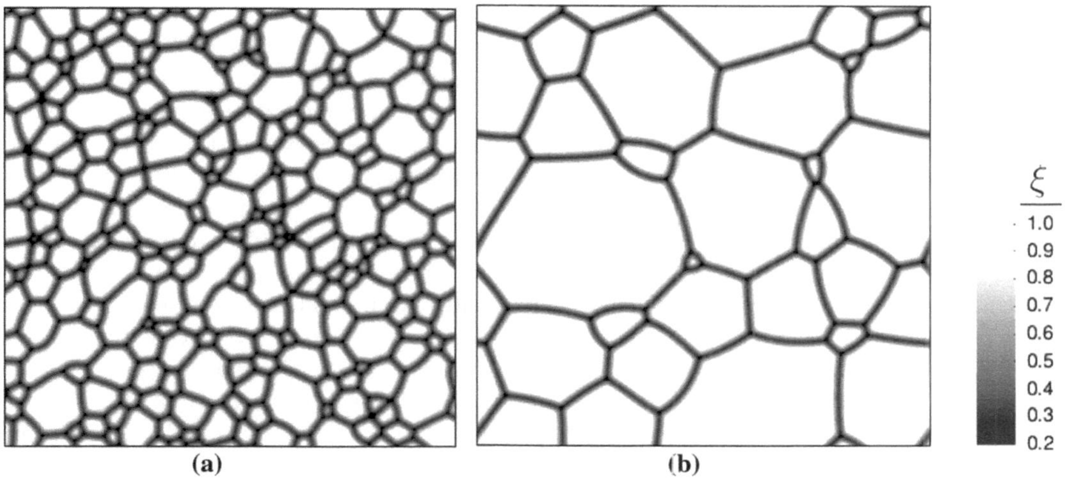

Fig. 1 Two snapshots of the polycrystalline microstructure evolution at time steps **a** $t = 125$ **b** $t = 1{,}750$. The *gray* contour indicates the distribution of ξ. Grain boundaries are highlighted by *darker* regions

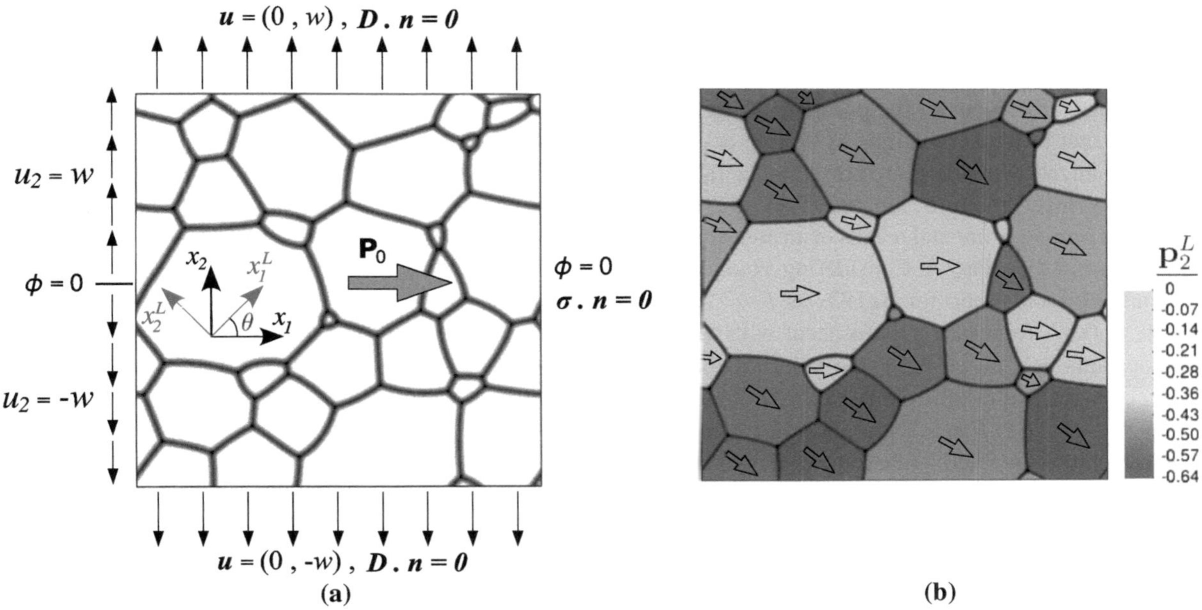

Fig. 2 **a** Computational model with boundary conditions; the initial polarization $\mathbf{p}_0$ is along the positive x_1 direction in the global coordinate system. The orientation of each grain is assigned randomly as $0° \leqslant \theta \leqslant 45°$ which indicates the angle between the global coordinate system $(x_1 - x_2)$ of the polycrystal and the local coordinate system $(x_1^L - x_2^L)$ of each grain. Same boundary conditions are also applied for the fracture simulation of the fine-grain structure. **b** Distribution of the initial polarization in each grain. The orientations of the initial polarization vectors are indicated with *arrows*

that with the horizontal initial polarization and choosing the grain orientations between 0° and 45°, ferroelastic switching becomes favorable with an acceptable mismatch between the grains, consequently leading to strong interactions between the microstructure and the crack propagation. For clarification purposes, the initial polarization in each grain is presented in Fig. 2b for the coarse-grain structure. As for the electrical boundary conditions, the electric potential on the left and right sides of the domain is set to 0. It is also assumed that the normal component of the electric displacement vanishes on other surfaces, i.e. $\mathbf{D} \cdot \mathbf{n} = 0$, representing the free-space. All the boundaries satisfy the free-polarization boundary condition.

For convenience, dimensionless variables are selected through the normalizations presented in Table 1. The constants are chosen to fit the behavior of single crystals of barium titanate ($BaTiO_3$) at room temperature, taking $c_0 = 1\,\text{GPa}$, a value for the spontaneous polarization of $p_0 = 0.26\,\text{C/m}^2$, the relative spontaneous strains $\varepsilon_a = -0.44\%$ along a-axis and $\varepsilon_c = 0.65\%$ along c-axis (Zhang and Bhattacharya 2005a; Wang et al. 2007). The domain wall scaling parameter is set to $a_0 = 3.7 \times 10^{-9} \frac{\text{Vm}^3}{\text{C}}$. The normalized scaling parameter of the domain wall energy a_0' is used to adjust the domain wall width in the computational domain. This parameter has to be chosen such that the variation of the polarization can be resolved by the discretization while domain walls remain sufficiently sharp relative to the other dimensions of the problem. These conditions are met in the simulations by setting $a_0' = 0.1$. The normalized parameters are presented in Table 2.

The intrinsic fracture toughness of $BaTiO_3$ is obtained from experimental results of an annealed sample as $K_c = 0.49\,\text{MPa}\sqrt{m}$. Considering Young's modulus and Poisson's ratio for $BaTiO_3$ as $E = 100\,\text{GPa}$ and $\nu = 0.37$, respectively (consistent with the elastic constants $c_i, i = 1, 2, 3$), the value of the critical energy release rate in plane-strain is obtained as $G_c = (1 - \nu^2)K_c^2/E = 2\ \text{J/m}^2$. The value of normalized critical energy release rate is then calculated as $G_c' = 4$. Using this value and the obtained distributions of ξ in Fig. 1a and b, the critical energy release rate is defined over both microstructures in Eq. (12).

The residual stiffness η_κ must be chosen as small as possible to avoid adding too much artificial stiffness and permittivity to the elements in the fracture zone, while preserving the non-singularity of the stiffness matrices used for the solution of finite element equations. We take $\eta_\kappa = 10^{-6}$ without any observed numerical instabilities in the simulations.

The value of the fracture regularization parameter κ is chosen based on parametric studies of the discretized surface energy (Bourdin et al. 2008; Amor et al. 2009). These numerical experiments indicate that setting $\kappa \sim h$ gives reasonable results, although the computed surface energy can be expected to be slightly overestimated. For the simulations of this paper, the regularization parameter is set to twice the element size as $\kappa = 2$, i.e. $h/\kappa = 0.5$. The remaining constants are chosen as follows: two tolerances $\delta_{ferro} = 10^{-3}$ and $\delta_{vfield} = 2 \times 10^{-3}$, the threshold to detect the fracture zone $\zeta = 2 \times 10^{-2}$ and the inverse of the mobilities $\mu_p = 1$ and $\mu_v = 15$. One hundred load steps are computed ($n = 100$) with load increments of $\Delta w^n = 3 \times 10^{-2}$. The normalized time step $\Delta t_m' = 10^{-2}$ leads to convergent and accurate solutions for the numerical time integration of the gradient flow equations in Eqs. (18) and (19).

With the selected parameters, the lengthscales of the model are obtained as follows. The width of the domain walls and grain boundaries is spanned with 4–6 elements in the simulations, corresponding to 2–3 nm, in the order of the experimentally measured values in tetragonal ferroelectric ceramics (Stemmer et al. 1995; Floquet et al. 1997; Tan and Shang 2002). The crack is smeared over 10 elements resulting in a width of 5 nm. The average grain sizes of the fine- and coarse-grain structures are obtained as 5 and 20 nm, respectively. Note that the physical grain size of the ferroelectric polycrystals is in the order of micrometers. In this micrometer scale, extremely fine meshes would be needed to model the nanoscale features such as the domain walls and grain boundaries, demanding massive parallel processing. However, the present configuration provides a proof of concept of the model and allows for a qualitative study. More physically realistic computations are on the way.

Figures 3 and 4 present the final computed crack paths (when the sample is splitted into two parts) in the fine- and coarse-grain structures, respectively. It is obvious in Fig. 3 that the crack propagates mainly along the grain boundaries, demonstrating the intergranular mode of fracture in the fine-grain structure. On the other hand, transgranular crack propagation is observed in Fig. 4 where the crack propagates through the grains. Experimental observations in ferroelectric

Table 2 Normalized parameters

a_1'	a_2'	a_3'	a_4'	a_5'	ε_0'	G_c'
−0.007	−0.009	0.018	0.0261	5	0.131	4
c_1'	c_2'	c_3'	b_1'	b_2'	b_3'	a_0'
185	111	74	1.4282	−0.185	0.8066	0.1

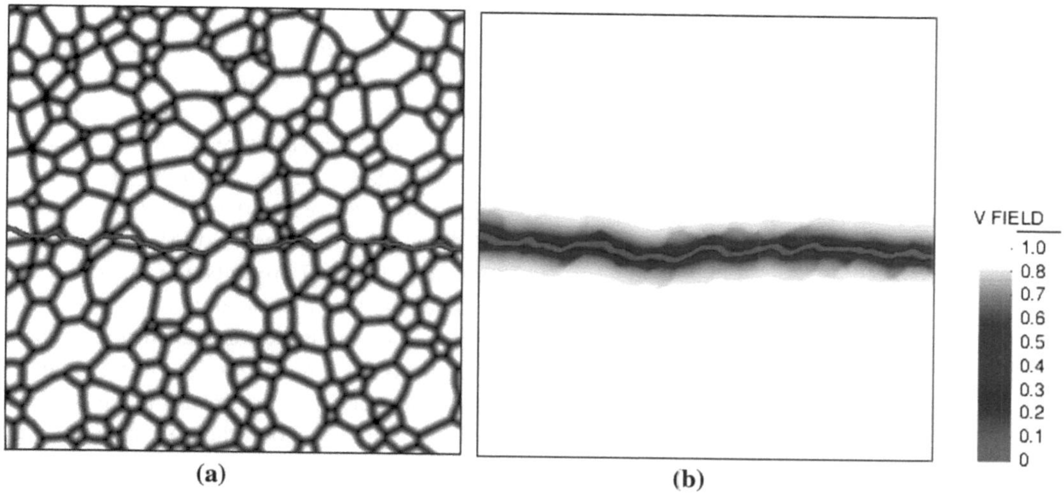

Fig. 3 Crack path in the fine-grain ferroelectric polycrystal **a** the points where $v = 0$ are depicted in *red* to indicate the crack position **b** field v representing the smeared crack

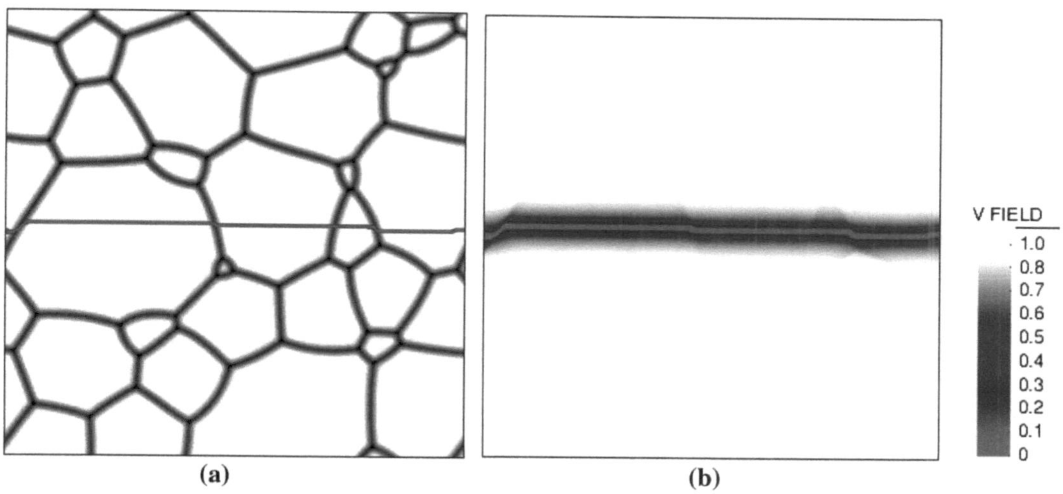

Fig. 4 Crack path in the coarse-grain ferroelectric polycrystal **a** the points where $v = 0$ are depicted in *red* to indicate the crack position **b** field v representing the smeared crack

polycrystals also show this transition from intergranular fracture for small grains to transgranular fracture in the case of large grains (Kim et al. 1990; Meschke et al. 1997).

To evaluate the effect of the fracture toughening mechanisms, the evolution of the normalized crack length (the second integral in Eq. (17) without the factor G_c^{poly}) is plotted in Fig. 5 as a function of the load step w. Both the fine- and coarse-grain graphs start from zero (intact model), and are almost identical until the cracks initiate at load step $w = 1.17$. After this point, the crack in the fine-grain structure starts to propagate abruptly, i.e. a small increase of the load leads to a big jump in the crack length. This is because after cutting the first grain, the crack starts to propagate faster along the grain boundaries with the lower fracture toughness. Interestingly, the abrupt propagation stops at load step $w = 1.2$ (point **a** in the fine-grain graph). This crack arrest can be easily understood by tracking the position of the crack tip in

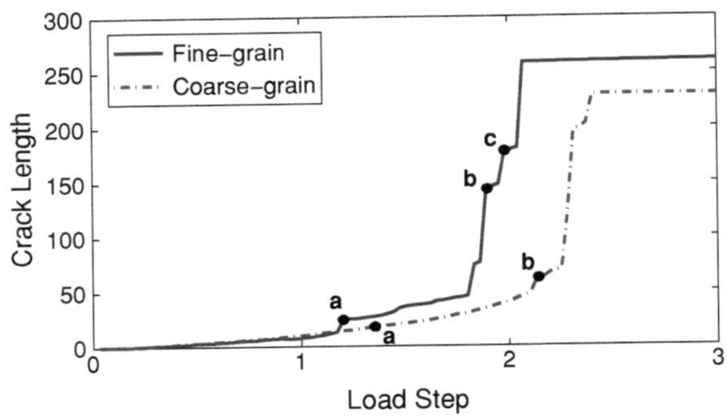

Fig. 5 Evolution of the normalized crack length as a function of the load step. The *letters*, marked in the graphs of the fine- and coarse-grain structures, correspond to the load steps shown in Figs. 6 and 7, respectively

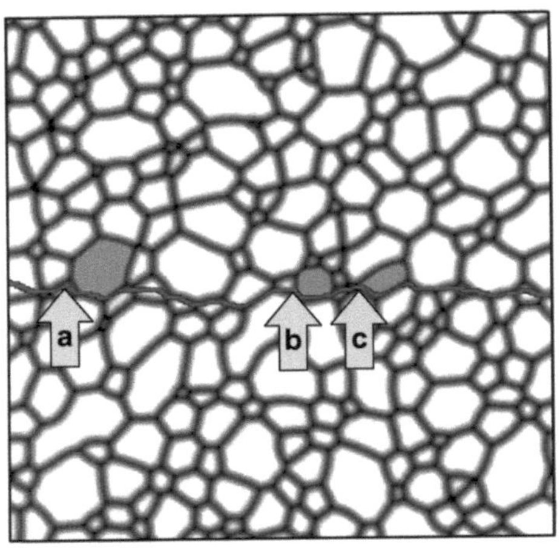

Fig. 6 Three steps of the crack propagation in the fine-grain structure. The *letters a–c* correspond to the load steps marked in the graph of the fine-grain structure in Fig. 5. In each load step, the position of the crack tip is indicated by the *yellow arrow* and the grain in front of the crack tip is highlighted with *green*. Crack deflection around the highlighted grains is obvious in this figure

this load step, as presented in Fig. 6. As the crack meets the first highlighted grain at load step **a**, it is forced to move around the grain since the fracture toughness of the grain interior is higher then the grain boundary, leading to crack deflection along the grain boundary. This phenomena is the so-called crack deflection toughening in intergranular mode of fracture (Faber and Evans 1983; Kueck et al. 2008). The crack shows this behavior in other load steps **b** and **c**, where the crack is arrested by the corresponding highlighted grains in Fig. 6.

Regarding the crack propagation in the coarse-grain microstructure, its crack-length graph in Fig. 5 shows a slower growth rate than the fine-grain structure. This is mainly due to the fact that, the transgranular mode of fracture in the coarse-grain structure allows the crack to interact with the ferroelectric domain microstructure inside the grains. Due to the high tensile stresses around the fracture zone, the polarization vectors start to rotate towards the vertical direction which leads to 90° ferroelastic domain switching. Figure 7 presents two snapshots of the evolution of the ferroelectric domain microstructure. The formation of a multiple-domain structure ahead of the crack tip is evident in Fig. 7a. The switched regions form twins ahead of the crack tip. Since the crystal's unit cell is longer along the polarization direction, as encoded in the Landau-Devonshire model, compressive stresses in the vertical (x_2) direction are induced by the vertical twins, leading to a strong toughening effect, already pointed out in the literature (Schneider 2007; Abdollahi and Arias 2011b). 90° ferroelastic domain switching is also favorable in the grains with a high misalingment since the polarization vectors in these grains can easily rotate towards the vertical direction by the tensile loading. This leads to the formation of multiple twins around the crack shielding the crack tip from the applied loading. As the load increases, these twins grow and additionally, a new set of twins nucleates in the unswitched regions, cf. Fig. 7a and b. The slow crack propagation periods, observed in the graph of Fig. 5, correspond to the load steps with a considerable expansion of 90° switched domains. Therefore, the main toughening mechanism in a coarse-grain ferroelectric polycrystal is the domain switching-induced toughening. This toughening mechanism is also reported in experiments of crack propagation in $BaTiO_3$ (Meschke et al. 1997, 2000; Fang et al. 2007). In contrast, 90° ferroelastic switching is less pronounced in the fine-grain structure since the crack propagates faster along the grain boundaries with a lower applied load. It is obvious in Fig. 5 that the crack length in the fine-grain structure reaches its maximum value sooner than the coarse-grain structure. This maximum level correspond to the point when the crack splits the model into two parts. Therefore the effective fracture toughness of the coarse-grain structure is higher than the fine-grain, which agrees with experiments (Meschke et al. 1997). Due to the crack deflection in the fine-grain structure, the final length of the crack is larger than that of the coarse-grain structure.

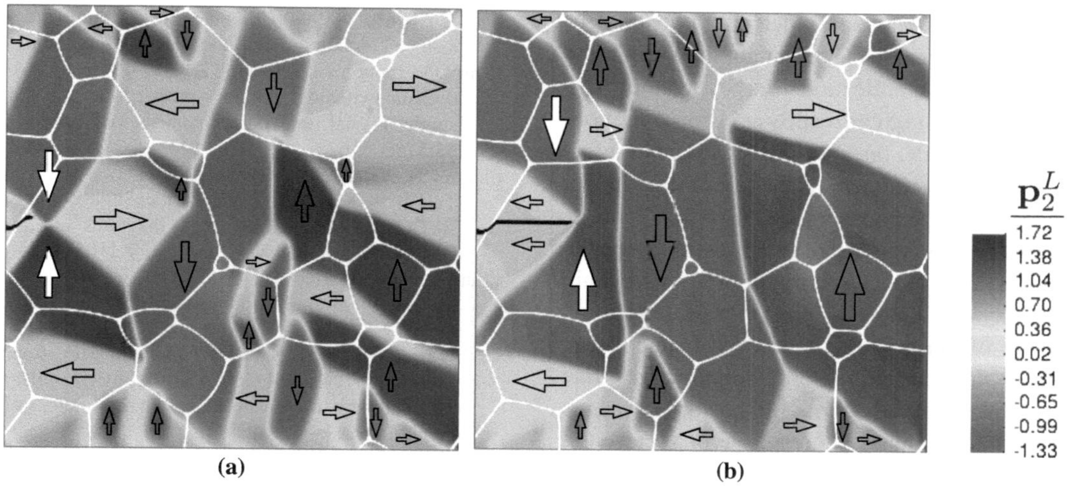

Fig. 7 Two snapshots of the evolution of the ferroelectric domain microstructure and the crack path at load steps **a** $w = 1.32$, **b** $w = 2.1$. These figures correspond to the load steps marked in the crack length graph of the coarse-grain structure in Fig. 5. The local *vertical* polarization $\mathbf{p}_2^L$ highlights the domain structure. Domain orientations are indicated with *arrows*, which are *bold white* for the twins ahead of the crack. The points where $v = 0$ are represented in *black* to show the crack position. Grain boundaries are highlighted by *white lines*

4 Conclusions

We perform, to the best of our knowledge, the first simulation of crack propagation in ferroelectric polycrystals with the goal of qualitatively evaluating the effect of the grain size and ferroelectric domain microstructures on the fracture process. For this purpose, we extend to polycrystals a phase-field model of fracture in ferroelectric single-crystals by incorporating the differential fracture toughness of the bulk and grain boundaries, and different crystal orientations of the grains. Simulation results show intergranular and transgranular crack propagations in fine- and coarse-grain structures, respectively. Crack deflection is observed as the main toughening mechanism in the intergranular mode of fracture in the fine-grain structure. On the other hand, as transgranular fracture is the dominant mode in coarse-grain microstructures, the crack interacts with the ferroelectric domains inside the grains. In particular, 90° ferroelastic domain switching results in a considerable fracture toughening effect. All these observations agree with experiment.

We also suggest that more work is needed for a quantitative evaluation of the combined effect of the polycrystalline microstructure and the ferroelectric domain microstructure on the fracture response of ferroelectric polycrystals. In particular, the relative physical size of the grains and ferroelectric domains should be considered for more realistic simulations. Besides the grain size, the model presents three lengthscales: (1) the width of the grain boundaries, (2) the width of the ferroelectric domain walls, and (3) the width of the smeared cracks. The first two lengthscales can be estimated experimentally (Stemmer et al. 1995; Floquet et al. 1997; Tan and Shang 2002). The latter is a numerical artifact, which could have a significant effect on the simulation results. A sensitivity analysis on the regularization parameter of the fracture phase-field relative to the other lengthscales of the problem would be necessary, as well as an extension to three dimensions. Although domain switching is claimed to be the main toughnening mechanism in coarse-grain ferroelectric polycrystals, the intrinsic anisotropy of the crystal in each grain, i.e. the different surface energy in different cleavage planes, is bound to have an effect as well. Besides, there is evidence that the properties of the grain boundaries depend on the crystallographic misorientation of the neighboring grains. Another important issue is the quantification and relative magnitude of the mobility parameters μ_p and μ_v, which can have an important effect on the resulting response. These topics are currently under investigation.

Finally, in the context of the multiscale paradigm, the model proposed here targets a mesoscale, half-way between atomistic models and phenomenological macro-scale models. Although, the bridging to the smaller and larger scale models is beyond the scope of the present paper, one can readily envision some connections. On one side, one can extract valuable information about the parameters of the phase-field model from atomistic calculations, possibly *ab initio*. Parameters susceptible of atomistic feed are those defining the Ginzburg-Laundau model, the mobility of the domain walls, the fracture toughness along cleavage planes and the structure of the grain boundaries, to name a few. On the other side, our ability to model polycrystalline structures and eventually three dimensional systems will allow us to simulate the response of representative volume elements in a micromechanical approach to extract effective properties for macroscopic phenomenological models.

Acknowledgments The authors gratefully acknowledge the support of the Ministerio de Ciencia e Innovaci6n (DPI2010-19145).

References

Abdollahi A, Arias I (2011a) Modeling and numerical simulation of crack propagation in piezoelectric and ferroelectric materials considering different crack face conditions (submitted for publication)
Abdollahi A, Arias I (2011b) Phase-field modeling of the coupled microstructure and fracture evolution in ferroelectric single crystals. Acta Mater 59:4733–4746
Abdollahi A, Arias I (2011c) Phase-field simulation of anisotropic crack propagation in ferroelectric single crystals: effect of microstructure on the fracture process. Model Simul Mater Sci Eng 19:074010
Amor H, Marigo JJ, Maurini C (2009) Regularized formulation of the variational brittle fracture with unilateral contact: numerical experiments. J Mech Phys Solids 57:1209–1229
Bourdin B, Francfort GA, Marigo JJ (2008) The variational approach to fracture. J Elast 91:5–148
Dadvand P, Rossi R, Onate E (2010) An object-oriented environment for developing finite element codes for multi-disciplinary applications. Arch Comput Methods Eng 17:253–297
DeSimone A, Kohn RV, Müller S, Otto F, Schäfer R (2001) Two-dimensional modeling of soft ferromagnetic films. Proc R Soc Lond A 457:2983–2991
Devonshire AF (1949) Theory of barium titanate. 1. Philos Mag 40:1040–1063
Devonshire AF (1951) Theory of barium titanate. 2. Philos Mag 42:1065–1079
Faber KT, Evans AG (1983) Intergranular crack-deflection toughening in silicon-carbide. J Am Ceram Soc 66:C94–C96
Fan D, Chen LQ (1997) Computer simulation of grain growth using a continuum field model. Acta Mater 45:611–622
Fang DN, Jiang YJ, Li S, Sun CT (2007) Interactions between domain switching and crack propagation in poled BaTiO3 single crystal under mechanical loading. Acta Mater 55:5758–5767
Floquet N, Valot CM, Mesnier MT, Niepce JC, Normand L, Thorel A, Kilaas R (1997) Ferroelectric domain walls in BaTiO3: fingerprints in XRPD diagrams and quantitative HRTEM image analysis. J Phys III 7:1105–1128
Francfort GA, Marigo JJ (1998) Revisiting brittle fracture as an energy minimization problem. J Mech Phys Solids 46:1319–1342
Grah M, Alzebdeh K, Sheng PY, Vaudin MD, Bowman KJ, Ostoja-Starzewski M (1996) Brittle-intergranular failure in 2D microstructures: experiments and computer simulations. Acta Mater 44:4003–4018
Griffith AA (1921) The phenomena of rupture and flow in solids. Philos Trans Royal Soc Lond A 221:163–198
Hackemann S, Pfeiffer W (2003) Domain switching in process zones of PZT: characterization by microdiffraction and fracture mechanical methods. J Eur Ceram Soc 23:141–151
Huber JE (2005) Micromechanical modelling of ferroelectrics. Curr Opin Solid State Mater Sci 9:100–106
Jones JL, Motahari SM, VarlioglU M, Lienert U, Bernier JV, Hoffman M, Uestuendag E (2007) Crack tip process zone domain switching in a soft lead zirconate titanate ceramic. Acta Mater 55:5538–5548
Kamlah M (2001) Ferroelectric and ferroelastic piezoceramics—modeling of electromechanical hysteresis phenomena. Cont Mech Therm 13:219–268
Kim SB, Kim DY, Kim JJ, Cho SH (1990) Effect of grain-size and poling on the fracture mode of lead zirconate titanate ceramics. J Am Ceram Soc 73:161–163
Kueck AM, Kim DK, Ramasse QM, De Jonghe LC, Ritchie RO (2008) Atomic-resolution imaging of the nanoscale origin of toughness in rare-earth doped SiC. Nano Lett 8:2935–2939
Kuna M (2010) Fracture mechanics of piezoelectric materials—where are we right now?. Eng Fract Mech 77:309–326
Landis CM (2003) On the fracture toughness of ferroelastic materials. J Mech Phys Solids 51(8):1347–1369
Li Q, Kuna M (2011) Evaluation of electromechanical fracture behavior by configurational forces in cracked ferroelectric polycrystals. Comput Mater Sci. doi:10.1016/j.commatsci.2011.01.050
Li W, Landis CM (2011) Nucleation and growth of domains near crack tips in single crystal ferroelectrics. Eng Fract Mech. doi:10.1016/j.engfracmech.2011.01.002
Meschke F, Kolleck A, Schneider GA (1997) R-curve behaviour of $BaTiO_3$ due to stress-induced ferroelastic domain switching. J Eur Ceram Soc 17:1143–1149
Meschke F, Raddatz O, Kolleck A, Schneider GA (2000) R-curve behavior and crack-closure stresses in barium titanate and (Mg, Y)-PSZ ceramics. J Am Ceram Soc 83:353–361
Schneider GA (2007) Influence of electric field and mechanical stresses on the fracture of ferroelectrics. Annu Rev Mater Res 37:491–538

Sheng JS, Landis CM (2007) Toughening due to domain switching in single crystal ferroelectric materials. Int J Fract 143:161–175

Shu YC, Bhattacharya K (2001) Domain patterns and macroscopic behaviour of ferroelectric materials. Phil Mag B 81:2021–2054

Song YC, Soh AK, Ni Y (2007) Phase field simulation of crack tip domain switching in ferroelectrics. J Phys D Appl Phys 40:1175–1182

Stemmer S, Streiffer SK, Ernst F, Ruhle M (1995) Atomistic structure of 90-degrees domain-walls in ferroelectric $PbTiO_3$ thin-films. Phil Mag A 71:713–724

Sukumar N, Srolovitz DJ, Baker TJ, Prevost JH (2003) Brittle fracture in polycrystalline microstructures with the extended finite element method. Int J Numer Methods Eng 56:2015–2037

Tan XL, Shang JK (2002) In-situ transmission electron microscopy study of electric-field-induced grain-boundary cracking in lead zirconate titanate. Phil Mag A 82:1463–1478

Verhoosel CV, Gutierrez MA (2009) Modelling inter- and transgranular fracture in piezoelectric polycrystals. Eng Fract Mech 76:742–760

Wang J, Landis CM (2006) Effects of in-plane electric fields on the toughening behavior of ferroelectric ceramics. J Mech Mater Struct 1:1075–1095

Wang J, Zhang TY (2007) Phase field simulations of polarization switching-induced toughening in ferroelectric ceramics. Acta Mater 55:2465–2477

Wang J, Zhang TY (2008) Phase field simulations of a permeable crack parallel to the original polarization direction in a ferroelectric mono-domain. Eng Fract Mech 75(17): 4886–4897. doi:10.1016/j.engfracmech.2008.06.025

Wang YL, Tagantsev AK, Damjanovic D, Setter N, Yarmarkin VK, Sokolov AI, Lukyanchuk IA (2007) Landau thermodynamic potential for BaTiO3. J Appl Phys 101(1–9):104,115

Xu BX, Schrade D, Gross D, Mueller R (2010) Fracture simulation of ferroelectrics based on the phase field continuum and a damage variable. Int J Fract 166:163–172

Yang L, Dayal K (2011) Microstructure and stray electric fields at surface cracks in ferroelectrics. Int J Fract. doi:10.1007/s10704-011-9670-2

Yang W, Zhu T (1998) Switch-toughening of ferroelectrics subjected to electric fields. J Mech Phys Solids 46:291–311

Zhang TY, Gao CF (2004) Fracture behaviors of piezoelectric materials. Theor Appl Fract Mech 41:339–379

Zhang W, Bhattacharya K (2005a) A computational model of ferroelectric domains. Part I: model formulation and domain switching. Acta Mater 53:185–198

Zhang W, Bhattacharya K (2005b) A computational model of ferroelectric domains. Part II: grain boundaries and defect pinning. Acta Mater 53:199–209

Zhu T, Yang W (1997) Toughness variation of ferroelectrics by polarization switch under non-uniform electric field. Acta Mater 45:4695–4702

Zhu T, Yang W (1999) Fatigue crack growth in ferroelectrics driven by cyclic electric loading. J Mech Phys Solids 47: 81–97

Int J Fract (2012) 174:17–27
DOI 10.1007/s10704-011-9670-2

ORIGINAL PAPER

Microstructure and stray electric fields at surface cracks in ferroelectrics

Lun Yang · **Kaushik Dayal**

Received: 6 September 2011 / Accepted: 20 December 2011 / Published online 17 January 2012

Abstract Ferroelectric perovskites are widely used in transducer, memory and optical applications due to their attractive electromechanical and optical properties. In these brittle materials, reliability and failure of devices is dominated by the behavior of cracks. The electromechanical coupling causes cracks to interact strongly with both mechanical as well as electrical fields. Additionally, cracks and domain patterns interact strongly with each other. Hence, an understanding of the electromechanics of cracks requires an accounting of all these interactions. In this work, we apply a real-space phase-field method to compute the stresses, domain patterns, and stray electric fields in the vicinity of a stationary crack, defined here as a geometric feature that causes large but bounded stress. We investigate the effects of charge compensation on the crack face, crack orientation with respect to the crystal lattice, and applied far-field stress and electric fields.

Keywords Ferroelectrics · Cracks · Phase-field simulation · Boundary element method · Domain patterns

L. Yang · K. Dayal (✉)
Carnegie Mellon University, Pittsburgh, PA, USA
e-mail: kaushik@cmu.edu

L. Yang
e-mail: luny@andrew.cmu.edu

1 Introduction

Ferroelectrics are widely used in sensors and actuators (Xu 1991; Uchino 1996), high-speed memory devices (Scott 2000), microwave circuits (Tagantsev et al. 2003; El-Naggar et al. 2006) and photonic switches (Dayal and Bhattacharya 2007b) due to their attractive electromechanical and optical properties. The reliability of ferroelectrics under the complex electrical and mechanical fields that occur in these device settings is an important issue. Due to the brittle nature of ferroelectrics, failure is dominated by the electromechanics of cracks. Therefore, predicting failure requires an understanding of the interactions between cracks, domain patterns, and stress and electric fields.

There have been numerous efforts to understand the behavior of cracks in ferroelectrics and related materials. Many of these have focused on domain switching near uniformly growing cracks, e.g. (Wang and Landis 2004; Landis 2003). More recently, phase-field models have been applied to stationary cracks (Xu et al. 2010, 2011; Li and Landis 2011; Wang and Zhang 2007) and also with sophisticated methods for growing cracks (Abdollahi and Arias 2011) that account elegantly for complex boundary conditions. Related experimental efforts have provided much insight into the behavior of cracks, e.g. some of the more notable include (Lynch 1998; Westram et al. 2007) and experiments on barium titanate single crystals using polarized light microscopy to observe individual domains (Jiang et al. 2009; Fang et al. 2007).

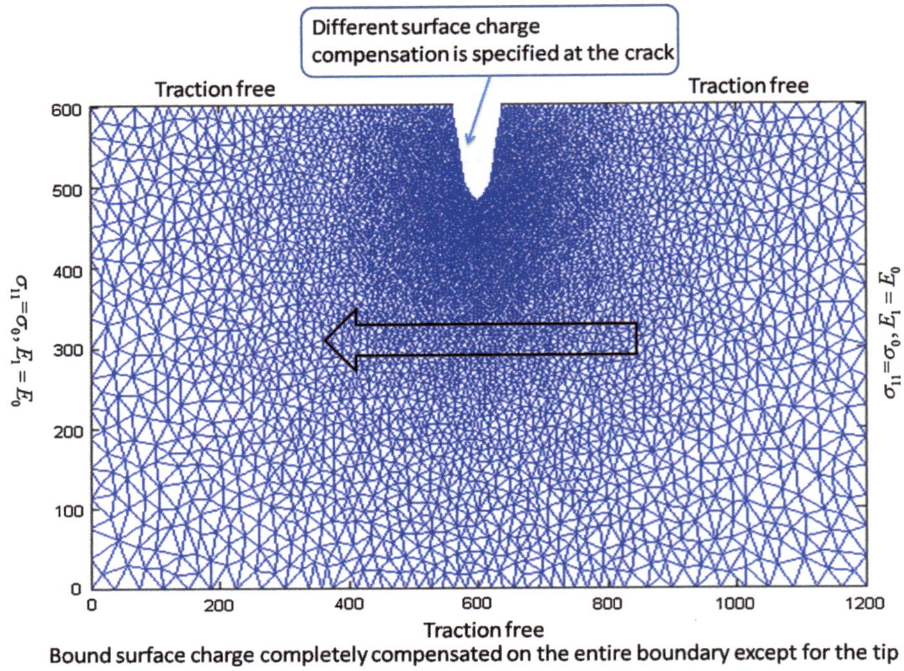

Fig. 1 Computational domain geometry, discretization and boundary conditions. Note that the *crack shape* in this figure is accurate, whereas the *jagged shapes* in subsequent figures are an artifact of the visualization software

Many of these efforts have been reviewed in Schneider (2007).

The key difference between this work and the theoretical/numerical efforts above is in our calculation of electric fields. In those works, simplifying assumptions on the surface charge and stray field are made, e.g. complete charge compensation at crack surface. While the problem of finding electric fields is posed on an infinite domain, the consequence of the simplifying assumptions is to reduce the problem to a finite domain corresponding to the ferroelectric specimen. In contrast, here we apply a recent boundary-element approach (Yang and Dayal 2011a; Dayal and Bhattacharya 2007a) to accurately and efficiently account for the stray fields by consistently transforming the infinite domain problem to an equivalent finite domain problem. A consequence of this approach is that we are able to compute stray fields as well as test the effect of different assumptions on crack-face charge compensation. In terms of the energy balance between stress, electric field, anisotropy energy, and boundary conditions that determines microstructure, we account accurately for the stray electric field contribution to the balance. In addition, stray fields provide an important experimental diagnostic due to the difficulty of directly observing microstructure in the interior. For example,

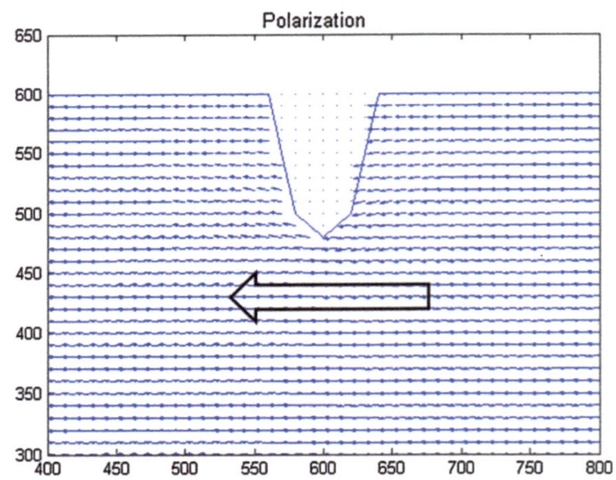

Fig. 2 Domain pattern with complete charge compensation on the crack surface

Schneider et al. (2003) and others use Kelvin Probe Microscopy to measure stray fields in ferroelectrics.

We apply the method to compute domain patterns, stresses, and electric fields in stationary cracks at the surface of a ferroelectric under different conditions. In our work, we define the crack as a geometric notch-shaped feature that leads to large but bounded stresses (Figs. 1, 2). In particular, we investigate the role of charge compensation on the crack face, crack orientation with respect to the crystal lattice, and applied far-field stress and electric fields.

The paper is organized as follows. The model formulation and the computational method are described in Sect. 2. The effect of varying charge compensation is described in Sect. 3. The effect of applied stress and electric field is described in Sect. 4. The effect of different lattice orientations is described in Sect. 5. We conclude in Sect. 6.

2 Phase-field model and solution method

The potential energy E of a ferroelectric Ω with strain field $\boldsymbol{\epsilon}$ and polarization field $\boldsymbol{p}$ is written, following Shu and Bhattacharya (2001):

$$E[\boldsymbol{\epsilon}, \boldsymbol{p}] = \int_{\Omega} \left[U(\nabla \boldsymbol{p}) + W(\boldsymbol{\epsilon}, \boldsymbol{p})\right] d\Omega + \frac{\epsilon_0}{2} \int_{\mathbb{R}^3} |\nabla\phi|^2 \, dV \tag{2.1}$$

with additional terms from loads if present. Here, U penalizes gradients in $\boldsymbol{p}$ and models the energy contributions from domain walls, and the stored (Devonshire/anisotropy) energy density W penalizes $\boldsymbol{p}$ and $\boldsymbol{\epsilon}$ deviating from the spontaneous polarization and strain. The last term is the energy contained in the electric field $\boldsymbol{E} = -\nabla\phi$, where ϕ is the electric potential obtained from Maxwell's equation:

$$\nabla \cdot (\boldsymbol{p} - \epsilon_0 \nabla\phi) = 0 \quad \text{over} \quad \mathbb{R}^3,$$

with appropriate boundary conditions. (2.2)

The first two terms in E from (2.1) are local to the body being considered, while the electrostatic contribution is over all space.

Domain patterns are obtained as local minimizers of E. These minimizers are numerically calculated by starting with assumed fields $\boldsymbol{\epsilon}$, $\boldsymbol{p}$ and evolving towards a minimum using a gradient flow. Unlike most previous works in phase-field modeling, e.g. Zhang and Bhattacharya (2005), Yang and Dayal (2010, 2011b), Li et al. (2010, 2011), we do not use the standard gradient flow based on the L_2 norm, but instead use the H_1 norm. The gradient flow of E is then:

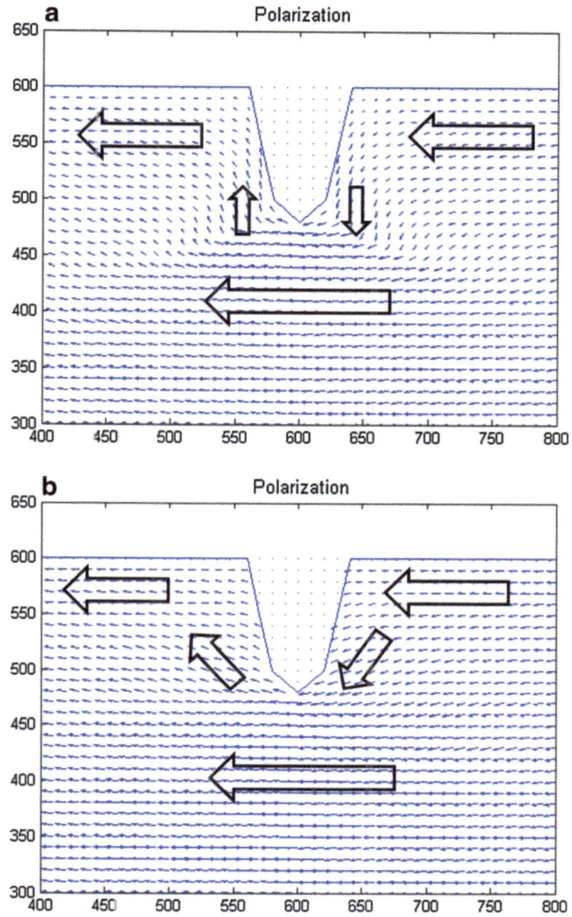

Fig. 3 Domain pattern with different charge compensation on the crack surface. **a** Uncompensated. **b** Partially compensated

$$\begin{aligned} \mu_1 \frac{dp_i}{dt} - \mu_2 \frac{d(p_{i,j})_{,j}}{dt} &= \left(\frac{\partial U}{\partial p_{i,j}}\right)_{,j} - \frac{\partial W}{\partial p_i} - \phi_{,i} \quad \text{over} \quad \Omega \\ \left(\frac{\partial W}{\partial \varepsilon_{ij}}\right)_{,j} &= 0 \quad \text{over} \quad \Omega \\ p_{i,i} - \epsilon_0 \phi_{,ii} &= 0 \quad \text{over} \quad \mathbb{R}^3 \end{aligned} \tag{2.3}$$

with μ_1, μ_2 both non-negative. The boundary conditions for the polarization evolution $(2.3)_1$ are $\nabla \boldsymbol{p} \cdot \boldsymbol{n} = 0$, where $\boldsymbol{n}$ is the outward normal to the boundary.

The boundary conditions for mechanical equilibrium $(2.3)_2$ are traction-free on the bottom and top faces of the domain, specified normal tractions on the sides of the domain, and traction-free on the crack faces (Fig. 2).

The boundary conditions for electrostatics $(2.3)_3$ are more varied. We recall that the bound surface due to

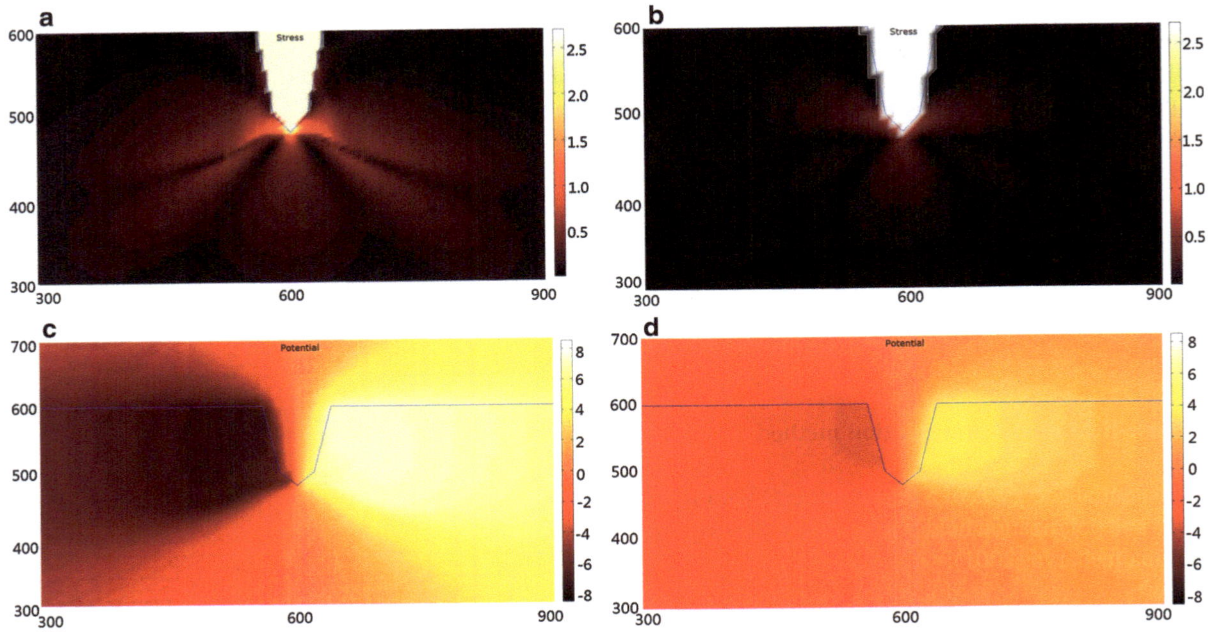

Fig. 4 Stress and potential fields with different charge compensation on the crack surface. **a** Von Mises stress with charge uncompensated. **b** Von Mises stress with charge partially compensated. **c** Potential with charge uncompensated. **d** Potential with charge partially compensated

polarization is $\sigma_b = \boldsymbol{p} \cdot \boldsymbol{n}$. Additionally, we can have free surface charge σ_f due to compensation from free charges in the ambient environment. The total charge on the boundary $\sigma_f + \sigma_b$. On the faces of the domain except along the crack, we assume complete compensation: $\sigma_f + \sigma_b = 0 \Leftrightarrow \sigma_f = -\boldsymbol{p} \cdot \boldsymbol{n}$. Along the crack faces, we use three models to examine the effect of charge compensation: first, we use $\sigma_f = -0.9\boldsymbol{p} \cdot \boldsymbol{n}$ that we term *complete charge compensation*; second, we use $\sigma_f = -0.5\boldsymbol{p} \cdot \boldsymbol{n}$ that we term *partial compensation*; and third, we use $\sigma_f = 0$ that we term *uncompensated*. In addition, we require that $\nabla\phi \to 0$ in the far-field and this automatically imposed by the Greens' functions of the numerical method (Yang and Dayal 2011a).

We start with an initial polarization field that we use to obtain the strain and electric potential from $(2.3)_{23}$. We then update the polarization using $(2.3)_1$. This process is repeated until convergence.

We use tetragonal barium titanate in two-dimensions as our model material, and use the energies given in Zhang and Bhattacharya (2005):

$$U(\nabla \boldsymbol{p}) = \frac{a_0}{2}\left(p_{1,1}^2 + p_{1,2}^2 + p_{2,1}^2 + p_{2,2}^2\right)$$

$$W(\boldsymbol{\epsilon}, \boldsymbol{p}) = \frac{a_1}{2}\left(p_1^2 + p_2^2\right) + \frac{a_2}{4}\left(p_1^4 + p_2^4\right) + \frac{a_3}{2}p_1^2 p_2^2 + \frac{a_4}{6}\left(p_1^6 + p_2^6\right) + \frac{a_5}{4}p_1^4 p_2^4 + \frac{1}{2}\left(\boldsymbol{\epsilon} - \boldsymbol{\epsilon}_T(\boldsymbol{p})\right) \cdot \mathbb{C} \cdot \left(\boldsymbol{\epsilon} - \boldsymbol{\epsilon}_T(\boldsymbol{p})\right) \quad (2.4)$$

The polynomial in the components of $\boldsymbol{p}$ in W provides a multi-well energy structure with energy wells that correspond the symmetry-related spontaneous polarizations, in this case aligned either along $(\pm 1, 0)$ or $(0, \pm 1)$. The stress-free strain $\boldsymbol{\epsilon}_T(\boldsymbol{p})$ is a function of polarization and represents the spontaneous strain associated with each value of polarization. Heuristically, the polynomial in W drives $\boldsymbol{p}$ to prefered spontaneous variants, and the stress-free strain then prefers the strain associated to that variant. The expression for $\boldsymbol{\epsilon}_T(\boldsymbol{p})$ is

$$\boldsymbol{\epsilon}_T = \begin{bmatrix} \frac{(c_1 b_1 - c_2 b_2)p_1^2 + (c_1 b_2 - c_2 b_1)p_2^2}{2(c_1^2 - c_2^2)} & \frac{b_3 p_1 p_2}{c_3} \\ \frac{b_3 p_1 p_2}{c_3} & \frac{(c_1 b_2 - c_2 b_1)p_1^2 + (c_1 b_1 - c_2 b_2)p_2^2}{2(c_1^2 - c_2^2)} \end{bmatrix} \quad (2.5)$$

The reason for the H_1-based gradient flow is that the standard L_2 gradient flow with explicit "time"-stepping requires small step size for CFL stability as a consequence of the linear diffusion-like term from $\left(\frac{\partial U}{\partial p_{i,j}}\right)_{,j}$. With the H_1 norm, we have a second derivative on the

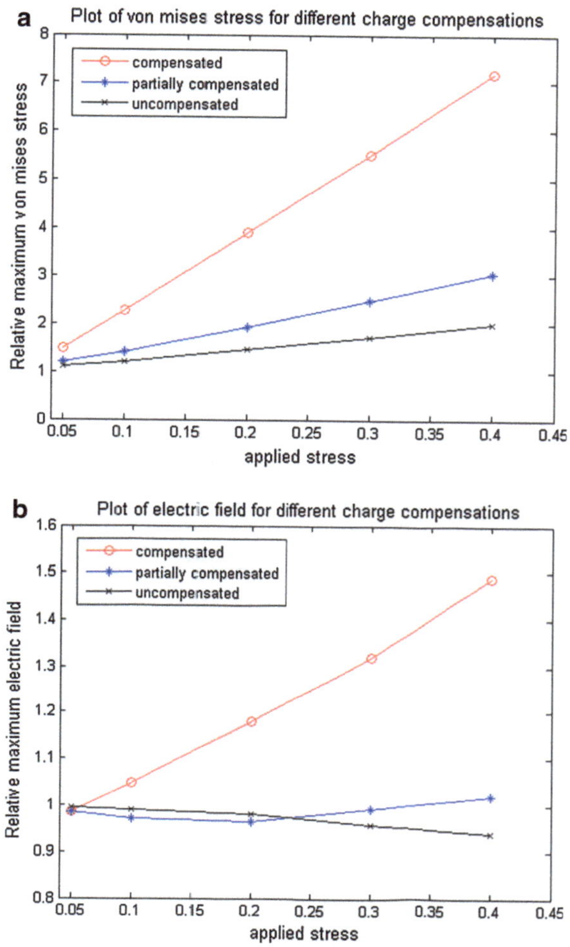

Fig. 5 Maximum stress and electric field with applied tensile stress σ_0. **a** Maximum Von Mises stress. **b** Maximum electric field

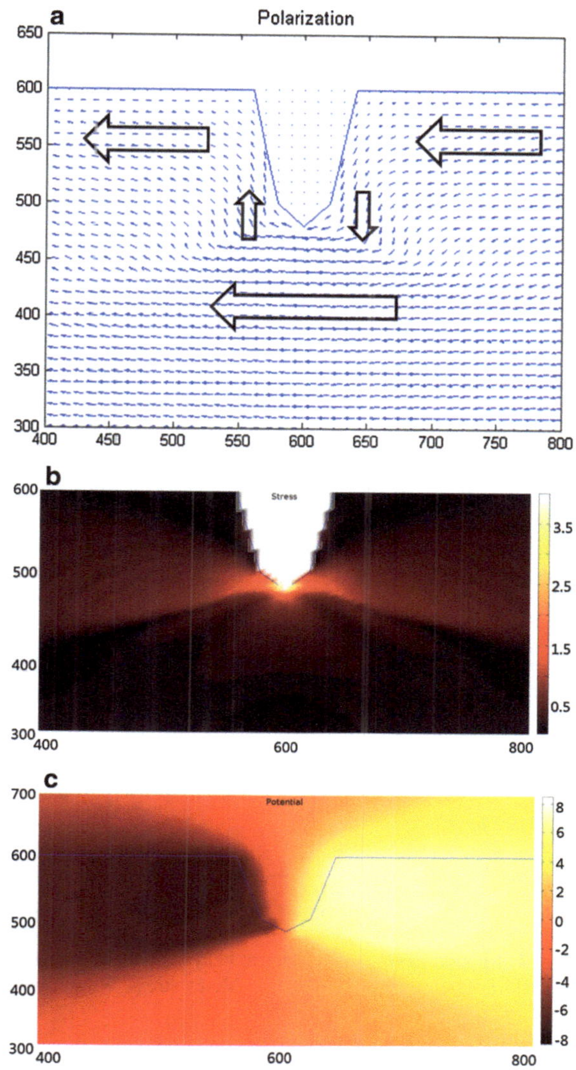

Fig. 6 Polarization, stress and potential without charge compensation on the crack surface when $\sigma_0 = +0.2$. **a** Polarization. **b** Stress. **c** Potential

left side that "cancels out" (in terms of CFL) the second derivative on the right in $(2.3)_1$. On taking the weak form and substituting the standard FEM ansatz in terms of shape functions, this new norm leads to a preconditioner that takes some account of the mesh. Ideally, we would set $\mu_1 = 0$ to maximize this cancellation, but this would lead to a seminorm with the boundary conditions $\nabla \boldsymbol{p} \cdot \hat{\boldsymbol{n}} = 0$, and the preconditioner would hence be singular. Hence, we use μ_1 as small as possible while retaining numerical stability of the preconditioner. The H_1 gradient flow leads to a wallclock time reduction of a factor of about 20.

The polarization update $(2.3)_1$ and mechanical equilibrium $(2.3)_2$ both use standard FEM with triangle elements and linear shape functions, and polarization and displacement are the primary variables respectively. The electrostatic equation $(2.3)_3$ needs attention as the electric fields are not confined to the ferroelectric but extend over all space. We use the boundary element approach to solve the infinite domain problem efficiently and accurately, as described in Dayal and Bhattacharya (2007a); Yang and Dayal (2011a). In particular, Yang and Dayal (2011a) describes a new iterative formulation that is particularly suited for nonlinear coupled problems such as arise here.

The generic geometry that we will consider in this paper is shown in Fig. 1 together with discretization,

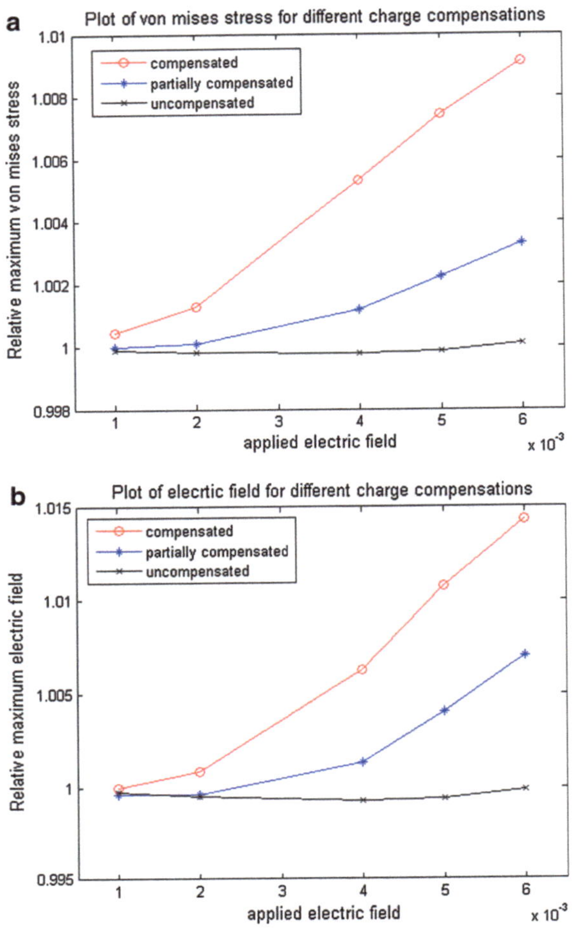

Fig. 7 Maximum stress and electric field with applied far-field electric field E_0. **a** Maximum Von Mises stress. **b** Maximum electric field

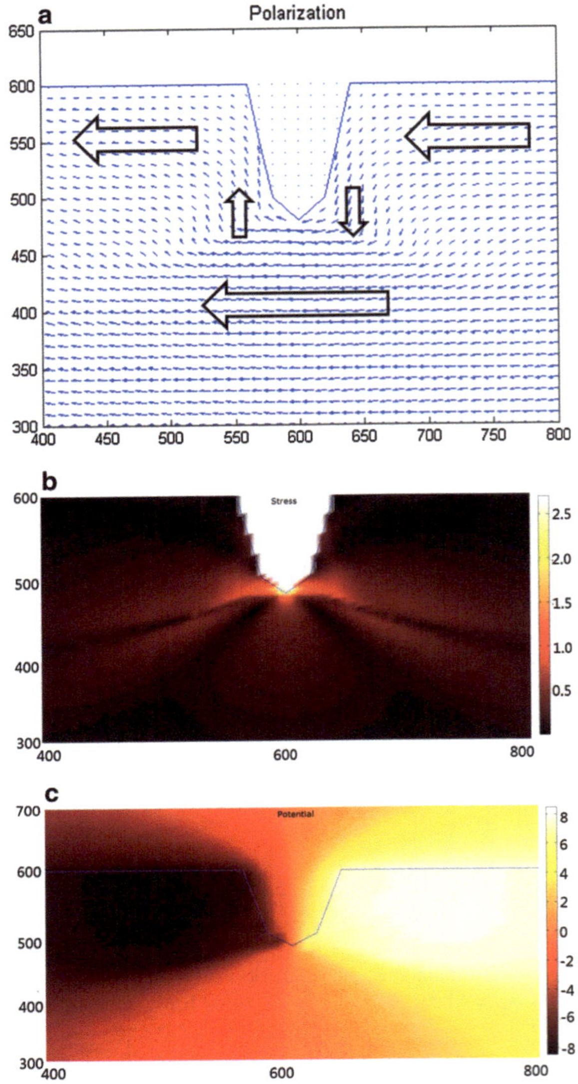

Fig. 8 Polarization, stress and potential without charge compensation on the crack surface when $E = 0.004$ **a** Polarization. **b** Stress. **c** Potential

boundary conditions and initial polarization. The mesh is refined near the crack tip. We note that the crack geometry is set up so that while there are large stresses, these are bounded and do not require special treatment.

3 Effect of charge compensation on the crack surface

The domain patterns near the crack tip involve a balance between elastic energy, the electrostatic energy, gradient energy, and the energy due to the polarization deviating from the preferred states. In a generic complex geometry, it is not possible to minimize each individually; for example, minimizing the electrostatic energy requires $\boldsymbol{p}$ to be tangential to the boundary everywhere to avoid bound surface charges, but this will obviously force the polarization to deviate from the preferred state. It is this particular aspect that we study in this section by varying the charge compensation on the crack face. This forces a competition between anisotropy energy, minimizing bound charges in the interior as well as on the surface with a complex geometry, and finally the complex stress fields due to the crack.

The complex orientation of the crack surface provides domain patterns that must balance all the energetic contributions. We recall that the surface bound

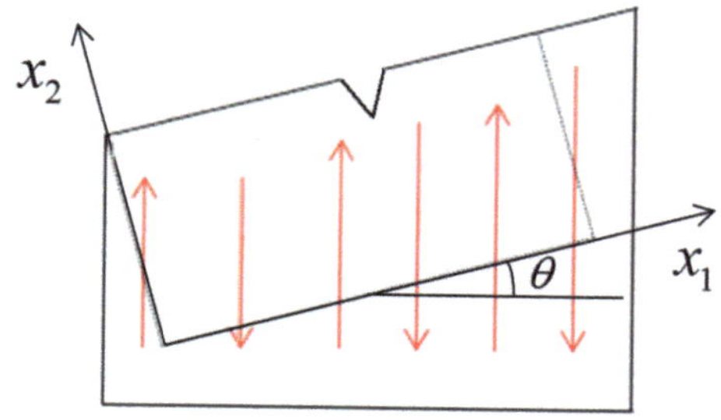

Fig. 9 Crack orientation relative to crystal lattice and computational domain

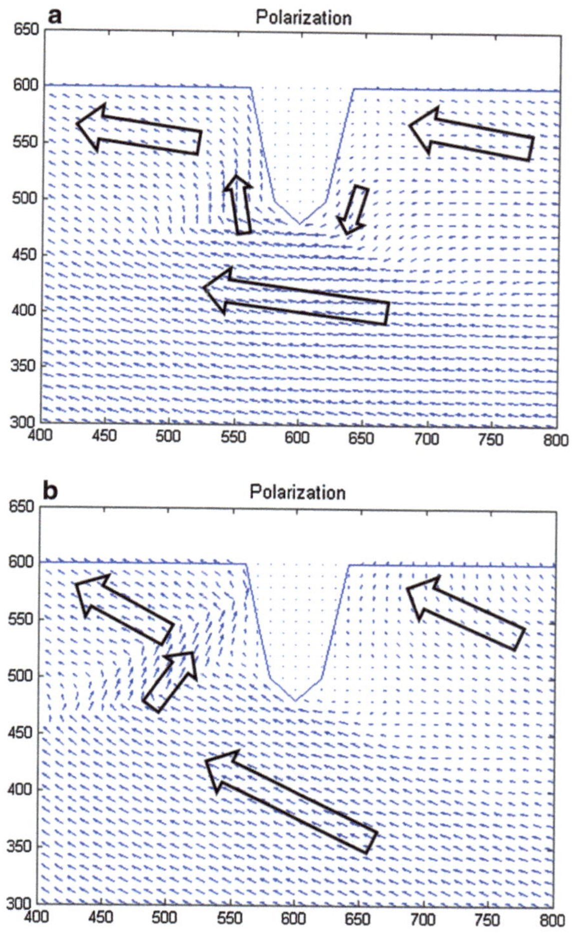

Fig. 10 Domain patterns without charge compensation on the crack surface with different lattice orientations **a** $\theta = 15°$. **b** $\theta = 30°$

charge due to a terminating polarization is $\boldsymbol{p} \cdot \boldsymbol{n}$. To examine the effect of charge compensation on the crack face, we use three models: complete charge compensation, partial compensation, uncompensated. And the top free surface excluding the crack, we assume the surface charge is completely compensated. In this section, we assume there is no applied far-field stress or electric field, that is $\sigma_0 = 0$ and $E_0 = 0$ in Fig. 1. We assume traction-free mechanical boundary conditions both at the top free surface and over the crack surface. In all cases, we use $\boldsymbol{p}$ aligned along the horizontal direction as initial condition.

Crack face electrical boundary conditions are usually assumed to be impermeable (no electric field in the gap), permeable (the crack does not perturb the electric field and is closed), and other cases between these extremes (Landis 2004). In a sense, the notion of crack permeability is related to the best approximate crack-face boundary condition to be used when solving electrostatics only on the body rather than all space. Because our approach solves for the external fields in all space, we do not use approximate crack-face conditions and it is not possible to make a direct comparison between our surface compensation approach and the approximate conditions. However, since we neither make the assumption that fields in the crack are zero, nor do we assume that the field is unperturbed by the crack, the closest case to our setting appears to be semi-permeable. A related issue is the fact that uncompensated charges on opposite crack faces cause mechanical tractions. The energy expression includes energetic contributions due to charge distribution. In the finite deformation context, this traction would appear simply by minimizing the energy with respect to displacement and polarization, as the minimization would include the energetic effect of moving the charges on the opposite sides of the crack face with respect to each other. In the small deformation approximation that we are using, this traction is not present and we have not added it separately.

In the completely compensated case, the domain pattern is fairly simple, Fig. 2. The initial condition is fairly low energy and there are only small energetic contributions both from the interior and the crack surface. Hence, the equilibrium state is quite close to starting guess.

We then examine the domain patterns with uncompensated charges. In this setting, the initial guess has small energetic contributions from the interior, but has large uncompensated bound charges on the crack surface due to normal termination of $\boldsymbol{p}$. Figure 3a shows the equilibrium domain pattern near the crack tip. The electrostatic energy dominates in this setting. As can be seen from Fig. 3a, $\boldsymbol{p}$ aligns tangential to the crack

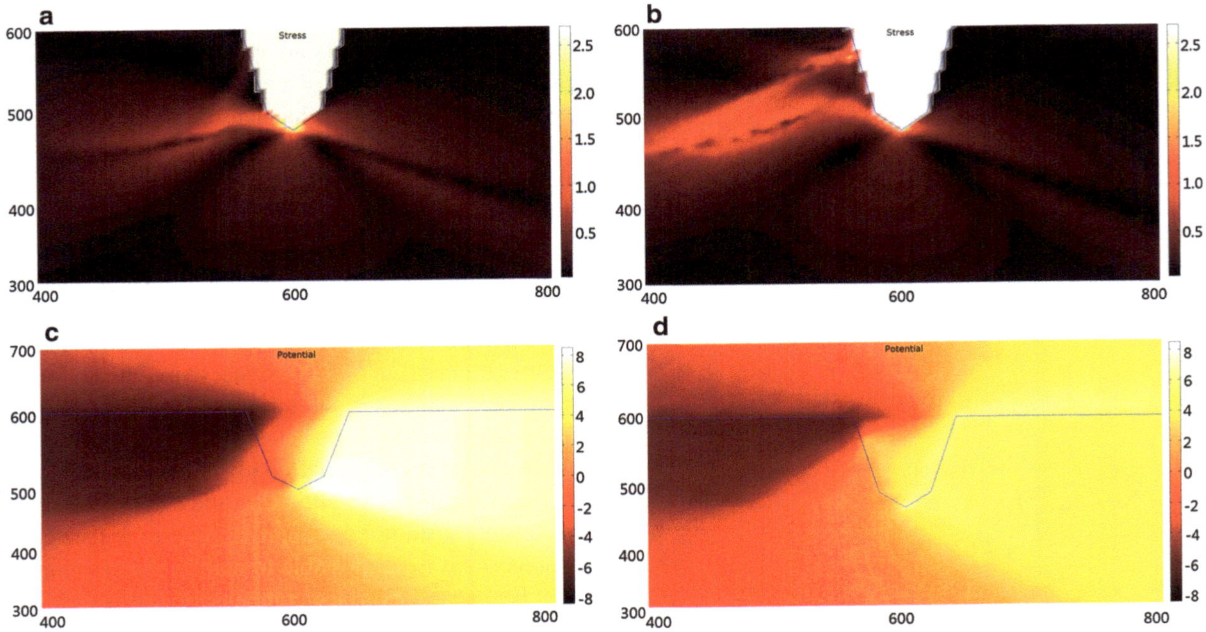

Fig. 11 Stress and electric potential fields without charge compensation on the crack surface with different lattice orientations. **a** Stress for $\theta = 15°$. **b** Stress for $\theta = 30°$. **c** Potential for $\theta = 15°$. **d** Potential for $\theta = 30°$

surface, and two 90° domain walls form on both sides of the crack. Also, the magnitude of $\boldsymbol{p}$ increases in the region directly below the crack due to the large electric fields and stresses, Fig. 4c and a respectively. As these latter figures show, the largest Von Mises stress and electric field are observed near the crack tip; besides these, relatively large stresses are observed in the region between the 90° walls, while large electric fields occur outside the material in the crack gap.

Finally, we examine the domain patterns with partial compensation. Polarization at the crack surface becomes partially tangential to balance the electrostatic energy with the anisotropy energy, Fig. 3b. Similar to the completely uncompensated case, the magnitude of $\boldsymbol{p}$ increases in the region directly below the crack tip. However, 90° domain walls are not observed. While the largest Von Mises stress and electric fields are observed near the crack tip, Fig. 4d and b respectively, the magnitudes of both quantities are much smaller than observed in the completely uncompensated setting.

4 Effect of far-field applied stress and electric field

In this section, we describe the effect of far-field applied normal stress and electric field, oriented as shown in Fig. 1.

First, we apply different values of far-field tensile stress σ_0 and focus on the consequent change in the maximum Von Mises stress in the ferroelectric and maximum electric field both inside and outside the ferroelectric. We do this with the three different charge compensations described in Sect. 3. Figure 5a and b show the normalized maximum Von Mises stress and the maximum electric field respectively for different values of σ_0. The normalization is with respect to the maximum Von Mises stress and electric field without applying any stress or electric field.

As expected, the maximum Von Mises stress increases with σ_0. The influence of different levels of charge compensation is also clearly felt as the increase is much larger when the charge is completely compensated, Fig. 5a. Similarly, the increase in the maximum electric field is also larger when the charge is completely compensated, Fig. 5b. The heuristic reason for this observation is that when the surface charge is not compensated, the formation of 90° domain walls and polarization switching provide a means for the specimen to accommodate the applied loads. Figure 6 shows the domain pattern, the Von Mises stress and the electric potential with no charge compensation when $\sigma_0 = 0.2$ (normalized). Both $\boldsymbol{p}$ and ϕ are comparable to Figs. 3a and 4c with no applied stress. However, the Von Mises stress

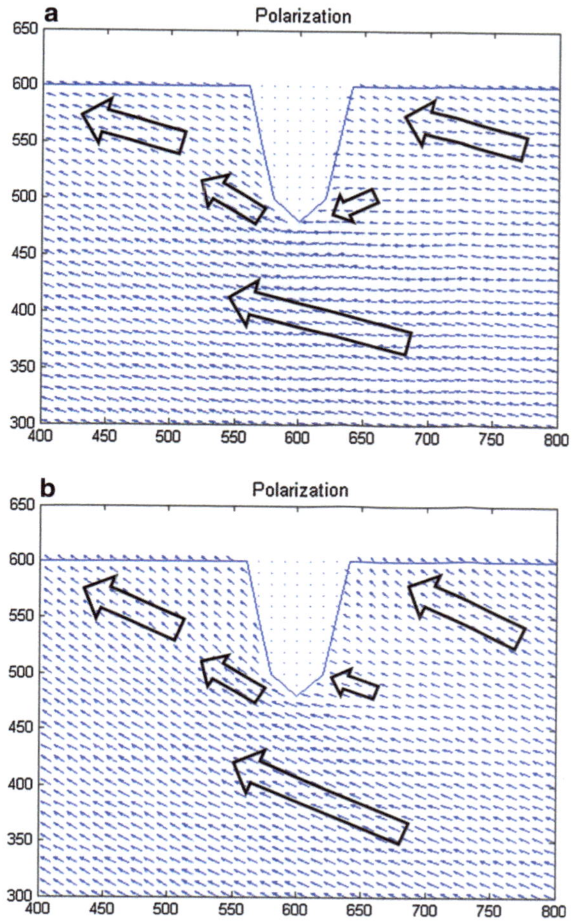

Fig. 12 Domain patterns with partial charge compensation on the crack surface with different lattice orientations. **a** $\theta = 15°$. **b** $\theta = 30°$

in Fig. 6b is significantly higher than in Fig. 4a. We also notice that the maximum electric field decreases slightly as σ_0 increases for $\sigma_0 \lesssim 0.2$ for both partially compensation and uncompensated, Fig. 5b. However, as σ_0 increases beyond 0.2, the electric field increases for partial compensation, while it continues to decrease if uncompensated.

Second, we apply a far-field electric field E_0 and study the effect on the maximum stress and electric field. Again, we compare the role of the three different levels of charge compensation. In general, E_0 induces only small changes in both the stress and electric field, Fig. 7a and b. This is consistent with the experimental observations of Sun et al. (2011) on a configuration that is very similar to our calculation. In particular, they too find that applied in-plane electric fields have little effect on the key features of the stray field. We also compare the polarization, stress and potential field with applied field, Fig. 8, with Figs. 3a, 4a and c that are computed in the absence of applied field and note that the changes are minor. However, different levels of charge compensation result in significantly different response of the stress and electric field when E_0 increases. Both the stress and electric field increase with increasing E_0 when the charge is completely or partially compensated, Fig. 7a and b. When uncompensated, both the stress and electric field initially slightly decrease and then slightly increase as E_0 increases; however, all these changes are small in comparison to the completely and partially compensated cases. Qualitatively, as with mechanical loading, the ferroelectric is able to accommodate applied electrical loading better when the electrostatic boundary conditions are less constraining as in the case of complete compensation.

5 Effect of crack orientation with respect to crystal lattice

In this section, we study the effect of the crack orientation with respect to the crystal lattice, by varying the orientation of the computational domain as in Fig. 9. In addition to the case of the crystal lattice oriented along the free surface ($\theta = 0°$) in previous sections, here we examine relative orientations of $\theta = 15°$ and $\theta = 30°$.

Figure 10a and b show the domain patterns for $\theta = 15°$ and $\theta = 30°$ respectively with no charge compensation. For $\theta = 15°$, we see two distinct 90° domain walls on the left side of the crack. In addition, the magnitude of $\boldsymbol{p}$ increases in the region beneath the crack. When $\theta = 30°$, the domain structure on the right side of the crack is indistinct, whereas prominent 90° domain walls form on the left side. Figure 11a, for $\theta = 15°$, and Fig. 4a, for $\theta = 0°$, show that the stress fields in these cases are similar. On increasing θ to 30°, Fig. 11b, we find a significant region of high stress developing on the left side of the crack. Figure 11c and d show the electric potential field for these cases.

Figure 12a and b show the domain patterns for $\theta = 15°$ and $\theta = 30°$ respectively with partial compensation. The polarization field adjusts slightly to the presence of the crack but does not form any distinct 90o domain walls in either case. Figure 13 shows that both stress and electric potential are much smaller compared to the case with no charge compensation, Fig. 11.

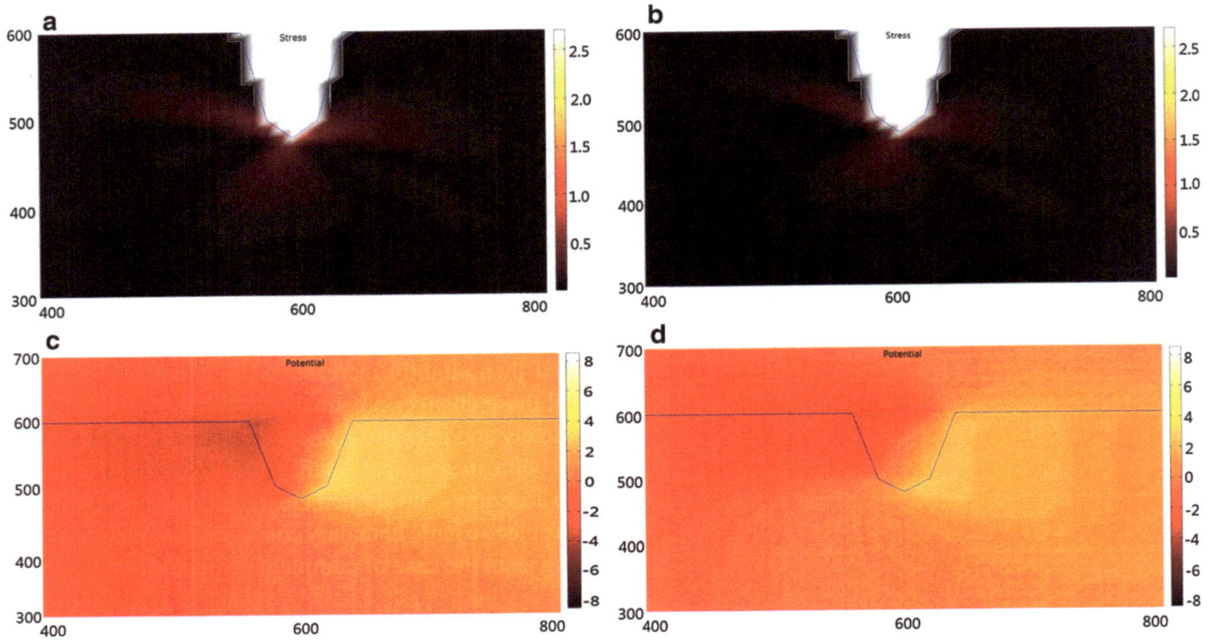

Fig. 13 Stress and electric potential fields with partial charge compensation on the crack surface with different lattice orientations. **a** Stress for $\theta = 15°$. **b** Stress for $\theta = 30°$. **c** Potential for $\theta = 15°$. **d** Potential for $\theta = 30°$

6 Conclusion

We have applied a real-space phase field method that can handle the stray electric fields to compute the microstructure in ferroelectrics with a crack with different levels of surface charge compensation. We have examined the effect the applied stress, electric field as well as crack orientation relative to crystal lattice on the domain patterns, stress fields, and stray electric fields in the crack vicinity. We find that the electrostatic energy dominates the energy balance if the surface charge on the crack face is not compensated. We find the formation of 90° domain walls on both sides of the crack in this situation. When far-field stress or electric field is applied, the microstructure is less sensitive to the applied field when the charge is completely compensated and hence the material has a larger response to applied field. In the same setting with no charge compensation, the boundary conditions allow some accommodation of the applied loads and hence the material response is smaller. In the case of different crack/lattice orientations, we find again that charge compensation plays a qualitatively similar role.

Acknowledgments This research was supported in part by the National Science Foundation through TeraGrid resources provided by the Pittsburgh Supercomputing Center. The real-space phase-field program described above is available at http://www.matforge.org. We thank Christoph Ortner for suggesting the H_1-based gradient flow.

References

Abdollahi A, Arias I (2011) Phase-field modeling of the coupled microstructure and fracture evolution in ferroelectric single crystals. Acta Mater 59(12):4733–4746

Dayal K, Bhattacharya K (2007a) A real-space non-local phase-field model of ferroelectric domain patterns in complex geometries. Acta Mater 55(6):1907–1917

Dayal K, Bhattacharya K (2007b) Active tuning of photonic device characteristics during operation by ferroelectric domain switching. J Appl Phys 102:064102

El-Naggar MY, Dayal K, Goodwin DG, Bhattacharya K (2006) Graded ferroelectric capacitors with robust temperature characteristics. J Appl Phys 100:114115

Fang D, Jiang Y, Li S, Sun CT (2007) Interactions between domain switching and crack propagation in poled batio3 single crystal under mechanical loading. Acta Mater 55(17):5758–5767

Jiang Y, Zhang Y, Liu B, Fang D (2009) Study on crack propagation in ferroelectric single crystal under electric loading. Acta Mater 57(5):1630–1638

Landis CM (2003) On the fracture toughness of ferroelastic materials. J Mech Phys Solids 51(8):1347–1369

Landis CM (2004) Energetically consistent boundary conditions for electromechanical fracture. Int J Solids Struct 41:6291–6315

Li LJ, Yang Y, Shu YC, Li JY (2010) Continuum theory and phase-field simulation of magnetoelectric effects in multiferroic bismuth ferrite. J Mech Phys Solids 58(10):1613–1627

Li LJ, Lei CH, Shu YC, Li JY (2011) Phase-field simulation of magnetoelastic couplings in ferromagnetic shape memory alloys. Acta Mater 59(7):2648–2655

Li W, Landis CM (2011) Nucleation and growth of domains near crack tips in single crystal ferroelectrics. Eng Fract Mech 78(7):1505–1513

Lynch CS (1998) Fracture of ferroelectric and relaxor electro-ceramics: influence of electric field1. Acta Mater 46(2):599–608

Schneider GA (2007) Influence of electric field and mechanical stresses on the fracture of ferroelectrics. Annu Rev Mater Res 37(1):491

Schneider GA, Felten F, McMeeking RM (2003) The electrical potential difference across cracks in pzt measured by kelvin probe microscopy and the implications for fracture. Acta Mater 51(8):2235–2241

Scott JF (2000) Ferroelectric memories. Springer, New York

Shu YC, Bhattacharya K (2001) Domain patterns and macroscopic behaviour of ferroelectric materials. Philos Mag Part B 81(12):2021–2054

Sun X, Su YJ, Gao KW, Guo LQ, Qiao LJ, Chu WY, Zhang TY (2011) Surface potential distribution in an indentation-pre-cracked batio3 single crystal. J Am Ceram Soc 94(12):4299–4304

Tagantsev AK, Sherman VO, Astafiev KF, Venkatesh J, Setter N (2003) Ferroelectric materials for microwave tunable applications. J Electroceram 11:5–66

Uchino K (1996) Piezoelectric actuators and ultrasonic motors. Kluwer, Dordrecht

Wang J, Landis CM (2004) On the fracture toughness of ferroelectric ceramics with electric field applied parallel to the crack front. Acta Mater 52(12):3435–3446

Wang J, Zhang TY (2007) Phase field simulations of polarization switching-induced toughening in ferroelectric ceramics. Acta Mater 55(7):2465–2477

Westram I, Oates WS, Lupascu DC, Rodel J, Lynch CS (2007) Mechanism of electric fatigue crack growth in lead zirconate titanate. Acta Mater 55(1):301–312

Xu BX, Schrade D, Gross D, Mueller R (2011) Fracture simulation of ferroelectrics based on the phase field continuum and a damage variable. Int J Fract 166(1):163–172

Xu BX, Schrade D, Gross D, Mueller R (2010) Phase field simulation of domain structures in cracked ferroelectrics. Int J Fract 165(2):163–173

Xu Y (1991) Ferroelectric materials and their applications. North-Holland

Yang L, Dayal K (2010) Formulation of phase-field energies for microstructure in complex crystal structures. Appl Phys Lett 96:081916

Yang L, Dayal K (2011a) A completely iterative method for the infinite domain electrostatic problem with nonlinear dielectric media. J Comput Phys 230(21):7821–7829

Yang L, Dayal K (2011b) Effect of lattice orientation, surface modulation, and applied fields on free-surface domain microstructure in ferroelectrics. Acta Mater 59(17):6594–6603

Zhang W, Bhattacharya K (2005) A computational model of ferroelectric domains. Part I: model formulation and domain switching. Acta Mater 53(1):185–198

Int J Fract (2012) 174:29–40
DOI 10.1007/s10704-012-9681-7

ORIGINAL PAPER

Double kink mechanisms for discrete dislocations in BCC crystals

M. P. Ariza · E. Tellechea · A. S. Menguiano · M. Ortiz

Received: 17 October 2011 / Accepted: 23 January 2012 / Published online: 11 February 2012

Abstract We present an application of the discrete dislocation theory to the characterization of the energetics of kinks in Mo, Ta and W body-centered cubic (BCC) crystals. The discrete dislocation calculations supply detailed predictions of formation and interaction energies for various double-kink formation and spreading mechanisms as a function of the geometry of the double kinks, including: the dependence of the formation energy of a double kink on its width; the energy of formation of a double kink on a screw dislocation containing a pre-existing double kink; and energy of formation of a double kink on a screw dislocation containing a pre-existing single kink. The computed interaction energies are expected to facilitates the nucleation of double kinks in close proximity to each other and to pre-existing kinks, thus promoting clustering of double kinks on screw segments and of 'daughter' double kinks ahead of 'mother' kinks. The predictions of the discrete dislocation theory are found to be in good agreement with the full atomistic calculations based on empirical interatomic potentials available in the literature.

Keywords Dislocations · Crystal plasticity · Double-kinks · BCC metals

M. P. Ariza (✉) · E. Tellechea · A. S. Menguiano
Escuela Técnica Superior de Ingeniería,
Universidad de Sevilla, 41092 Sevilla, Spain
e-mail: mpariza@us.es

E. Tellechea
e-mail: etellechea@us.es

A. S. Menguiano
e-mail: adrisanmen@us.es

M. Ortiz
Division of Engineering and Applied Science,
California Institute of Technology, Pasadena, CA 91125, USA
e-mail: ortiz@aero.caltech.edu

1 Introduction

The thermally activated double kink nucleation mechanism has attracted considerable attention over the years (Duesbery 1983; Tang et al. 1998; Cai et al. 2000; Yang and Moriarty 2001). This interest arises from the observation that, in body-centered cubic (BCC) metals, the motion of $\frac{a}{2}\langle 111\rangle$ screw dislocations is controlled by the nucleation and spreading of double kinks. The particular core structure of screw dislocations in BCC metals is accepted to be the underlying cause of this behavior (Vitek 1976). Some models of double-kink dynamics and BCC metal plasticity are based on linear-elastic dislocation theory (Hirth and Lothe 1968; Seeger 1981). Atomistic simulations based on empirical interatomic potentials or on lattice Green's functions have also been carried out in order to study the structure and energetics of the double kink configurations in BCC crystals (cf., e.g., Yang et al. 2001). In addition, dislocations mobility at low temperature in tantalum has been studied using a mesoscopic approach (Tang et al. 1998). In order to overcome the sample-size limitations of dislocation dynamics models, Cai et al. (2000, 2001) have modeled dislocation motion in silicon and molybdenum using a kinetic Monte Carlo scheme.

Discrete dislocation dynamics is a widely used computational tool for assessing the plasticity of single crystals. A number of approaches are commonly adopted in constructing discrete dislocation models. These approaches may roughly be grouped into line-tracking based methods (e.g. Zbib et al. 2002; Ghoniem et al. 2000; Madec et al. 2002) and non-tracking methods such as the phase-field and related models (Wang et al. 2001, 2004; Koslowski et al. 2002). Line-tracking methods require the discretization of dislocation lines and the subsequent computation of the long-range elastic interactions between pairs of dislocation segments. Special rules need to be set forth in order to account for topological transitions arising from dislocation reactions, annihilation, and others, and in order to regularize the logarithmically divergent self-interaction of the segments. A common additional limitation of linear elastic dislocation dynamics models arises from their reliance on isotropic elasticity, which neglects the elastic anisotropy of the crystals. Furthermore, the models fail to account for a dislocation-core structure and a minimum separation between slip-planes. Dislocation nucleation is often accounted for by means of ad hoc criteria and mobility laws require to be prescribed a priori.

The discrete dislocation theory of Ariza and Ortiz employed here overcomes many of the aforementioned limitations of linear elastic dislocation dynamics models. Thus, by acknowledging the discreteness of the lattice from the outset, the theory is endowed with several desirable features, including the elimination of unphysical singularities and the need for core cutoffs; fully anisotropic discrete lattice elasticity; an atomistic dislocation core structure; and topological transitions that are automatically accounted for without the need for any explicit rules of interaction. The computational advantages that are derived from the discrete dislocation theory of Ariza and Ortiz are also noteworthy. Thus, an appeal to algebraic topology greatly facilitates book-keeping in numerical calculations. Additionally, the discrete Fourier transform (DFT) enables the construction of fast and efficient computational schemes for large systems (Ramasubramaniam et al. 2007).

In this paper, we present an application of the discrete dislocation theory of Ariza and Ortiz to

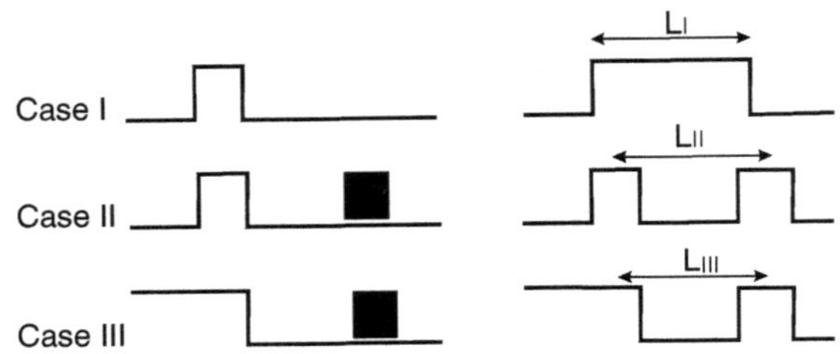

Fig. 1 Schematic representation of the transitions considered in this study

the characterization of the energetics of kinks in Mo, Ta and W BCC crystals. For definiteness, we restrict attention to the following transitions (Fig. 1):

- Case I: Nucleation of a double kink at an arbitrary position on an infinite screw dislocation and subsequent separation of each kink.
- Case II: Nucleation of a double kink at a certain position away from a preexisting double kink on an infinite screw dislocation.
- Case III: Nucleation of a double kink at a certain position away from a preexisting kink in an infinite screw dislocation.

In all cases, the transitions are effected by the addition of elementary loops, termed loopons by Shenoy et al. (1999). More generally, it is possible to transition between two arbitrary dislocation densities by means of a sequence of 'flips', each flip consisting on the addition or subtraction of an elementary loopon. For completeness, in Sect. 2 we begin by briefly summarizing the discrete dislocation theory that provides the basis for subsequent calculations. A study of the three configurations defined above is presented in Sect. 3. The discrete dislocation calculations supply detailed predictions of formation and interaction energies for various double-kink formation and spreading mechanisms as a function of the geometry of the double kinks. These predictions are found to be in good agreement with the full atomistic calculations based on empirical interatomic potentials available in the literature.

2 The lattice complex of BCC crystals

This section summarizes the general theory of discrete dislocations in crystal lattices, and its specialization to body-centered cubic (BCC) crystals, as developed in Ariza and Ortiz (2005), Ramasubramaniam et al. (2007).

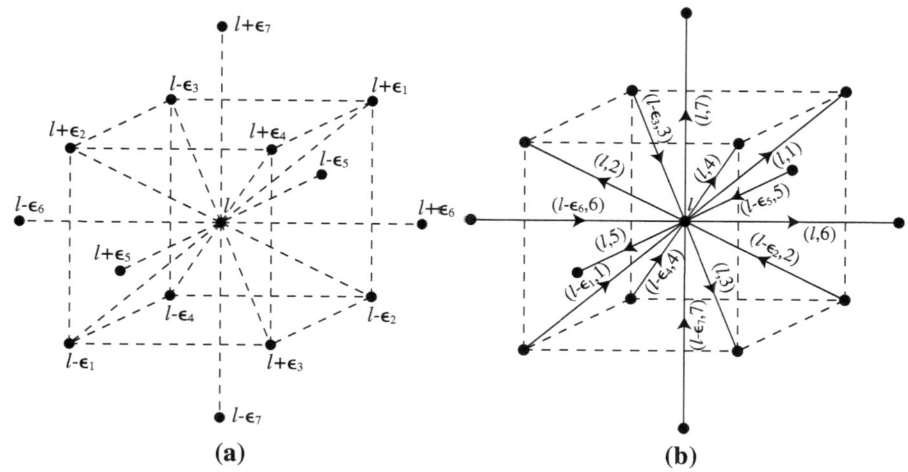

Fig. 2 CW-complex representation of the body-centered cubic lattice and complex lattice indexing scheme. **a** Atoms or 0-cells. **b** Atomic bonds or 1-cells

2.1 The cell complex of the BCC crystal lattice

Following Ariza and Ortiz, we regard the BCC lattice as a *cell-complex* $\mathcal{C}$, i.e., as a collection of cells of different dimensions equipped with discrete differential operators and a discrete integral. In particular, the BCC complex is three-dimensional and consists of: atoms, or 0-cells; atomic bonds, or 1-cells; atomic faces, or 2-cells; and tetrahedral cells, or 3-cells (Figs. 2, 3, 4). The BCC Bravais lattice is generated by the basis $\{(-a/2, a/2, a/2), (a/2, -a/2, a/2), (a/2, a/2, -a/2)\}$. The lattice complex representation of the BCC lattice is shown in Figs. 2, 3 and 4, where $\epsilon_1 = (1,0,0), \epsilon_2 = (0,1,0), \epsilon_3 = (0,0,1), \epsilon_4 = (1,1,1), \epsilon_5 = (0,1,1), \epsilon_6 = (1,0,1)$ and $\epsilon_7 = (1,1,0)$. Each element of the three-dimensional complex Bravais lattice can be indexed by means of three integers $\boldsymbol{l} \equiv (l^1, l^2, l^3) \in \mathbb{Z}^3$ and a label $\alpha \in \{1, \dots, N\}$ that designates which of the N simple Bravais sublattices the element belongs to.

The lattice cell complex supplies the support for defining functions, or forms, of different dimensions. Thus, a form of dimension p assigns a vector to each cell of dimension p of the lattice. As we shall see, forms provide the vehicle for describing the behavior of the BCC lattice, including its displacements, eigendeformations and dislocation densities. In addition, the description of lattice defects such as dislocations, and the formulation of the attendant equilibrium equations, is greatly facilitated by defining discrete differential operators over the lattice complex. To this end, we note that the BCC lattice complex defined in (4) is a simplicial complex and, hence, it can be endowed with the standard simplicial differential structure (Munkres 1984). To this end, we begin by orienting all cells (Figs. 2, 3, 4). Specifically, we recall that a p-simplex is defined as a collection $\sigma_p = \{v_0, \dots, v_p\}$ of vertices and that a simplex is oriented by choosing an ordering $\sigma_p = [v_0, \dots, v_p]$ of its vertices. Two simplices are then said to have the same orientation if the ordering of their vertices differs by an even permutation, and to have the opposite orientation if it differs by an odd permutation. The $(p-q)$-faces of a p-simplex σ_p are defined by removing q vertices from σ_p. An orientation of σ_p induces an orientation of its faces simply by keeping the vertices of the faces in the same order. Suppose that ω is a 0-form defined over the atoms and let $\sigma_1 = [v_a, v_b]$ be an ordered atomic bond defined by atoms a and b (cf. Fig. 5). Then, the differential $d\omega(\sigma_1)$ of ω at σ_1 is

$$d\omega(\sigma_1) = \omega(v_b) - \omega(v_a). \tag{1}$$

Suppose now that ω is a 1-form defined over the atomic bonds and let $\sigma_2 = \{v_a, v_b, v_c\}$ be a triangular cell whose boundary contains the atomic bonds $\sigma_{ab} = \{v_a, v_b\}, \sigma_{bc} = \{v_b, v_c\}$ and $\sigma_{ca} = \{v_c, v_a\}$ (cf. Fig. 5). Then, the differential $d\omega(\sigma_2)$ of ω at σ_2 is

$$d\omega(\sigma_2) = \pm\omega(\sigma_{ab}) \pm \omega(\sigma_{cb}) \pm \omega(\sigma_{ca}), \tag{2}$$

where the $+$ sign is assigned to those 1-cells on the right-hand side that are consistently oriented with σ_2 and the $-$ sign is assigned otherwise. The differential of a 2-form is defined likewise. Thus, the differential operator maps p-forms, defined over p-cells, to

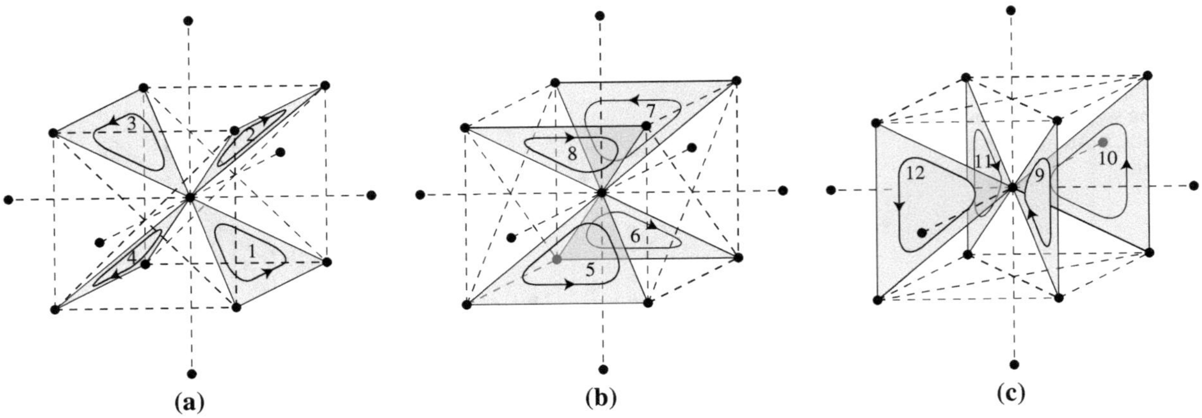

Fig. 3 CW-complex representation of the body-centered cubic lattice and complex lattice indexing scheme. Elementary faces or 2-cells

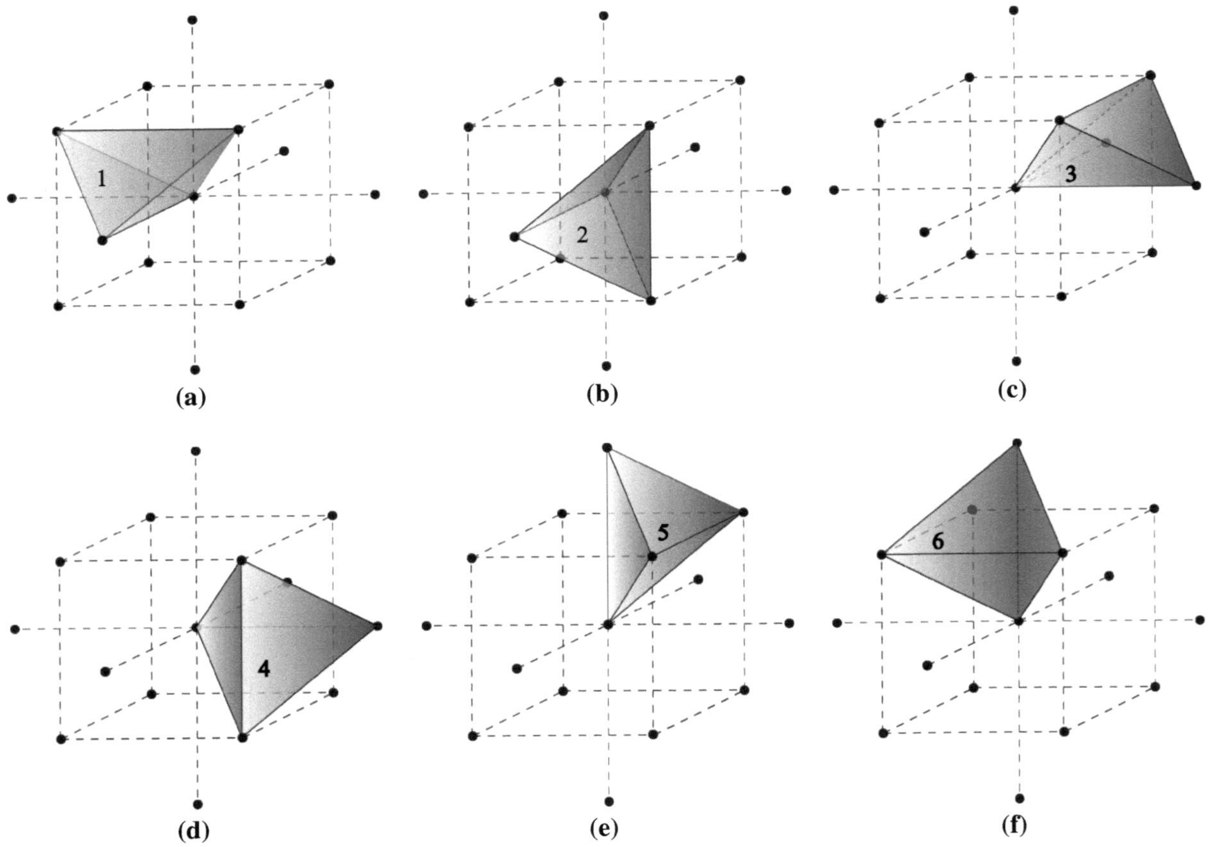

Fig. 4 CW-complex representation of the body-centered cubic lattice and complex lattice indexing scheme. Elementary volumes or 3-cells

$(p+1)$-forms, defined over $(p+1)$-cells. The discrete differential operators thus defined may be regarded as the discrete counterparts of the familiar grad, curl and div of vector calculus. In particular: the differential of 0-forms is the discrete counterpart of the grad operator; the differential of 1-forms is the discrete counterpart

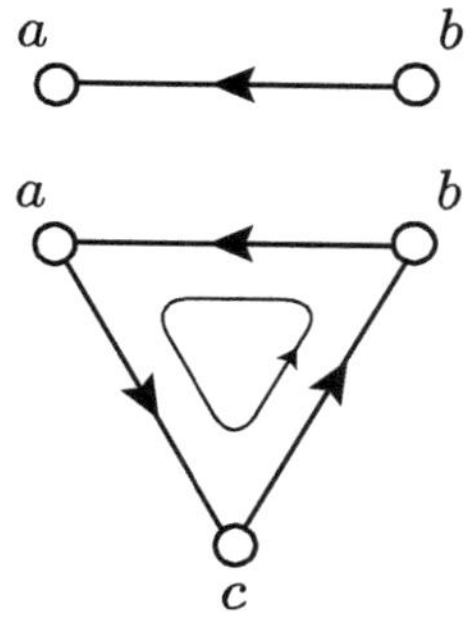

Fig. 5 Diagram for the definition of the discrete differential operators of BCC crystals

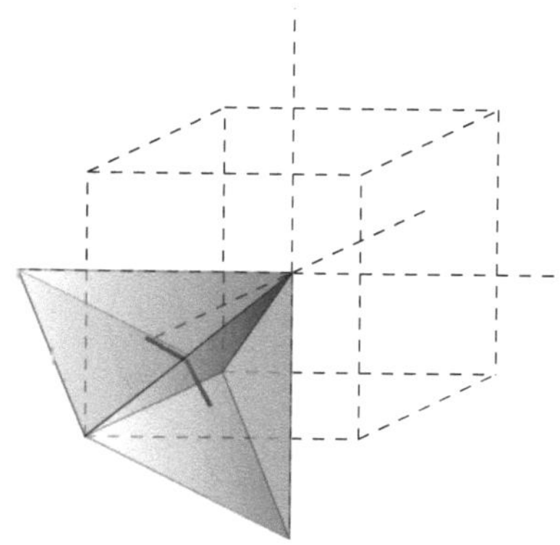

Fig. 6 Pair of elementary dual segments

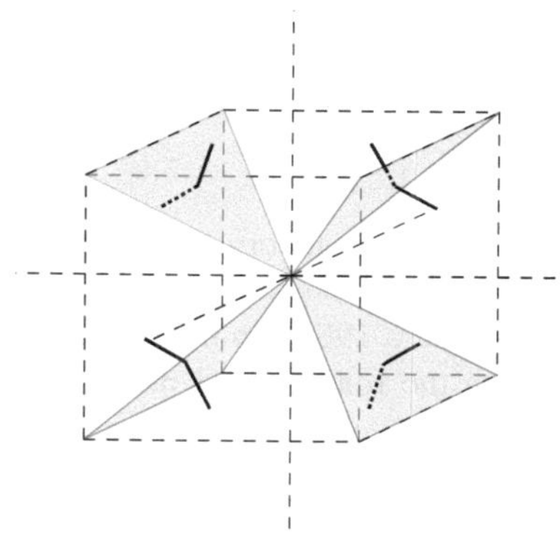

Fig. 7 Elementary segments of the dual lattice of BCC crystals

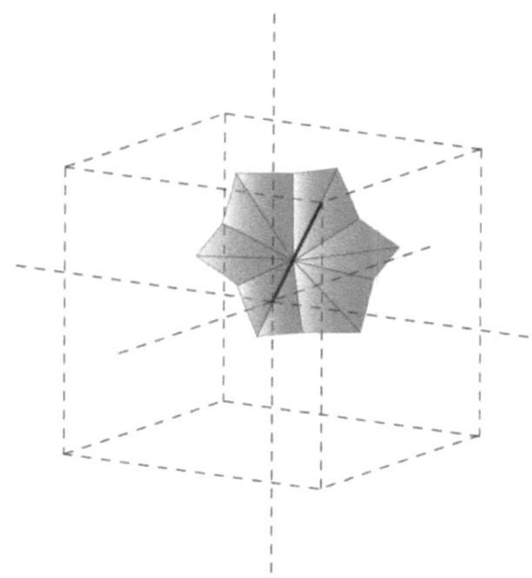

Fig. 8 Group of elementary faces that constitute the dual cell to an atomic bond of the BCC lattice

of the curl operator; and the differential of 2-forms is the discrete counterpart of the div operator from vector calculus. It is readily verified from the definition of the discrete differential operators that

$$d^2 = 0, \tag{3}$$

which is the discrete counterpart of the identities curl ∘ grad = 0 and div ∘ curl = 0. In applications, we shall additionally resort to the dual complex of the crystal lattice for representing discrete dislocation lines.

We recall that the p-cells of the dual lattice complex may be identified with the $(n - p)$-cells of the primal lattice complex (Munkres 1984). For simplicial complexes, the barycentric block decomposition (Munkres 1984) supplies a particular geometric representation of the dual lattice complex.

The block dual to each 3-cell of a cell-complex C is its barycenter. The closed block dual to any 2-cell consists of the two line segments joining the barycenter of this face with the barycenters of the two 3-cells having the former cell as a face (Figs. 6, 7). The block dual to any 1-cell is in general a non-flat region with flat faces and straight edges, that intersects the former cell at its barycenter. Each of these faces has three vertexes, the barycenter of the 1-cell, the barycenter of one of the 2-cells having the 1-simplex as an edge and the barycenter of the 3-cell having the 2-cell as a face (Fig. 8).

Likewise, elementary dual segments of the same type are translations of each other, have the same complement of neighbors and are arranged as simple Bravais lattices. The dual block to any 0-cell of C consists of the volume surrounded by the dual to the 1-cells having the former 0-cell in common. In Fig. 9 we have plotted the dual to a 0-cell whose edges are elementary dual segments.

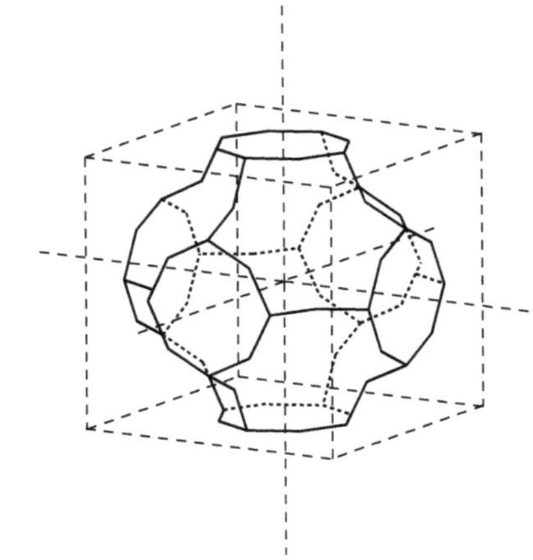

Fig. 9 Elementary dual segments that form the edges of the dual to a 0-cell of the BCC lattice

2.2 Eigendeformation theory of discrete lattice dislocations

It is possible to fashion a theory of discrete dislocations in crystals from the classical theory of eigendeformations (cf., e.g., Mura 1987). In the present setting, the theory rests on the fundamental property of crystals that certain uniform deformations leave the crystal lattice unchanged and, hence, should cost no energy. The entire class of lattice-invariant deformations is characterized by a classical theorem of Ericksen (1979) as the set of unimodular affine mappings with integer lattice coordinates. An energy that satisfies this property by construction is

$$E(u,\beta) = \frac{1}{2}\langle B(du-\beta),(du-\beta)\rangle \equiv \frac{1}{2}\sum_{e_1\in E_1(C)}\sum_{e'_1\in E_1(C)} \langle B(e_1,e'_1)(du(e_1)-\beta(e_1)),(du(e'_1)-\beta(e'_1))\rangle \tag{4}$$

where the sums take place over the atomic bonds of the crystal lattice and: $u(e_0)$ is the atomic displacement of atom e_0; $du(e_1)$ is the deformation of atomic bond e_1; $\beta(e_1)$ is the eigendeformation at bond e_1; and $B(e_1, e'_1)$ are bond-wise force constants. In (4), the local values $\beta(e_1)$ of the eigendeformation field are constrained to defining lattice-invariant deformations. By this restriction and the form of the energy (4), uniform lattice-invariant deformations du cost no energy, as desired. We note that, owing to the discrete nature of the set of lattice-invariant deformations, the energy (4) is strongly nonlinear. In particular, the reduce energy

$$E(u) = \inf_{\beta} E(u,\beta) \tag{5}$$

is piecewise quadratic with zero-energy wells at all uniform lattice-invariant deformations.

The primary slip systems of a BCC lattice are tabulated in Table 1. The atomic bond (Fig. 2b) were chosen so as to contain the $\frac{1}{2}\langle 111\rangle$ slip directions and 110 slip planes. We note from (4) that energy vanishes if the eigendeformations are compatible, i.e., if $\beta = dv$ for some atomic displacement field v. Indeed, in that case the energy is minimized for $u = v$ and the minimum energy is zero. Hence, the energy at equilibrium, or stored energy, i.e., the energy that remains stored in the crystal when the displacement field is equilibrated, can only depend on the degree of incompatibility of the eigendeformations. A measure of that incompatibility is provided by the discrete dislocation density

$$\alpha = d\beta, \tag{6}$$

which may be regarded as the discrete curl of the eigendeformations. Thus, α provides a discrete counterpart of Nye's dislocation density tensor field (Nye 1953). Since the eigendeformations β are defined on the atomic bonds, it follows that the discrete dislocations of BCC crystals are defined on the triangular cells of the lattice, i.e., the discrete dislocation density α assigns a Burgers vector to every 2-cell of the BCC lattice. From the fundamental property (3) of the discrete differential operator if follows that

$$d\alpha = 0, \tag{7}$$

i.e., the net Burgers vector of the discrete dislocation density must vanish. This condition is the discrete counterpart of the classical divergence-free property of the Nye's dislocation density tensor field, which may in turn be regarded as a conservation of Burgers vector property. A basis for the free-abelian group of the discrete dislocation densities generated by the eigendeformations β is shown in Fig. 10. In this figure, the elementary loops, or loopons (Shenoy et al. 1999), are represented as cycles of edges in the dual lattice cell complex (cf. Sect. 2). The loopons represent the dislocation density corresponding to unit slip on a single cartesian and diagonal primary edges. An arbitrary discrete dislocation density may be obtained through an integer linear combination of dislocation densities associated to elementary loops (Fig. 10). In particular, it follows that it is possible to transition between two arbitrary dislocation densities by means of a sequence

Table 1 BCC slip systems in Schmid and Boas' nomenclature. The unit normal to the slip plane is denoted by m and the unit vector along the Burgers vector by s

s. s.	A2	A3	A6	B2	B4	B5
$\sqrt{3}s$	$[\bar{1}11]$	$[\bar{1}11]$	$[\bar{1}11]$	$[111]$	$[111]$	$[111]$
$\sqrt{2}m$	$(0\bar{1}1)$	(101)	(110)	$(0\bar{1}1)$	$(\bar{1}01)$	$(\bar{1}10)$
s. s.	C1	C3	C5	D1	D4	D6
$\sqrt{3}s$	$[11\bar{1}]$	$[11\bar{1}]$	$[11\bar{1}]$	$[1\bar{1}1]$	$[1\bar{1}1]$	$[1\bar{1}1]$
$\sqrt{2}m$	(011)	(101)	$(\bar{1}10)$	(011)	$(\bar{1}01)$	(110)

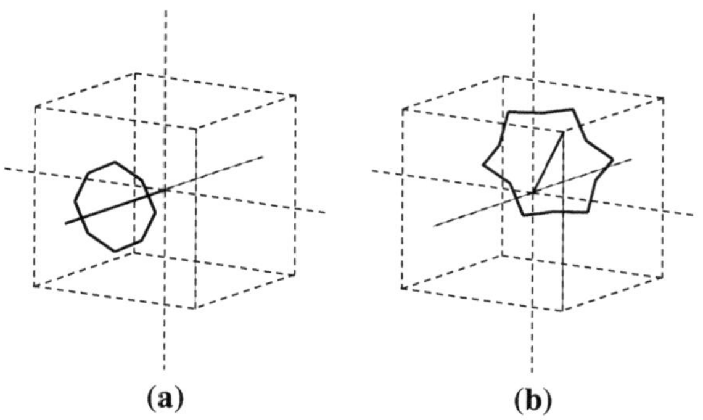

Fig. 10 Elementary dislocation loops, or 'loopons', in BCC crystals. **a** Cartesian loopon. **b** Diagonal loopon

of 'flips', each flip consisting on the addition or subtraction of an elementary loopon.

A theorem of Ariza and Ortiz (2005) shows that perfect lattices, including the graphene lattice (Ariza and Ortiz 2010), possess a Helmholtz–Hodge decomposition. By this discrete Helmholtz–Hodge decomposition, it follows, in particular, that $\alpha = 0$ if and only if $\beta = dv$ for some displacement field v, i.e., if and only if the eigendeformations are compatible. Thus, the discrete dislocation density does indeed provide a measure of the incompatibility of the eigendeformations. It also follows from the discrete Helmholtz–Hodge decomposition that α is determined by β up to an arbitrary displacement field. From these properties it may be shown (Ariza and Ortiz 2005)

$$\inf_u E(u, \beta) = E(\alpha), \tag{8}$$

i.e., that the stored energy of a crystal may be written as a function of the discrete dislocation density. By the quadratic dependence of the energy (4) on the displacement field it follows that the stored energy must be of the form

$$\begin{aligned} E(\alpha) &= \frac{1}{2} \sum_{l \in \mathbb{Z}^3} \sum_{l' \in \mathbb{Z}^3} \langle \Gamma(l - l')\alpha(l'), \alpha(l) \rangle \\ &\equiv \frac{1}{2} \langle \Gamma * \alpha, \alpha \rangle, \end{aligned} \tag{9}$$

where the kernel $\Gamma(l)$ gives the interaction energy between Burgers vectors (Table 1) at the origin and at position l on the simple Bravais lattice defined by the triangular cells (Fig. 3). Explicit expressions for the kernel Γ in terms of the force constants of the lattice are given in Ariza and Ortiz (2010). Again we note that, despite is harmonic appearance, the stored energy (9) is rendered strongly nonlinear by the constraint that the local Burgers vectors $\alpha(l)$ must be integer linear combinations of the basic Burgers vectors (Table 1). This strong nonlinearity renders the determination of low-energy dislocation structures mathematically challenging.

The DFT (cf., e.g., Babuska et al. 1960; Ariza and Ortiz 2005) provides a computationally convenient alternative representation of the stored energy (9). In order to define this representation, we begin by grouping cells of the same dimension by type (Figs. 2, 3, 4). Thus, cells of the same type are translations of each other and have the same complement of neighbors, or environment. According to this definition, a BCC crystal has one type of atoms, seven types of elementary segments, along each of the cube directions and diagonal directions, twelve types of elementary areas and six types of elementary volumes. The fundamental property of cells of the same type is that they are arranged as simple Bravais lattices (Figs. 2, 3, 4). Thus, the atoms of BCC crystals define one simple Bravais lattices, the atomic bonds define seven simple Bravais lattices, the triangular cells define twelve simple Bravais lattices, and the tetrahedral cells define six simple Bravais lattices. We recall that the DFT of a function f defined on a simple Bravais lattice $\mathbb{Z}^n$ is

$$\hat{f}(\theta) = \sum_{l \in \mathbb{Z}^n} f(l) \mathrm{e}^{-i\theta \cdot l}, \tag{10}$$

where the angle variables θ range over $[-\pi, \pi]^n$. The DFT admits an inverse given by

$$f(l) = \frac{1}{(2\pi)^n} \int_{[-\pi,\pi]^n} \hat{f}(\theta) \mathrm{e}^{i\theta \cdot l} \, d\theta, \tag{11}$$

and has properties similar to those of the Fourier transform, including a discrete Parseval identity and and a discrete convolution theorem. In the DFT representation, the stored energy (9) takes the form

$$E(\alpha) = \frac{1}{(2\pi)^3} \int_{[-\pi,\pi]^3} \langle \hat{\Gamma}(\theta)\hat{\alpha}(\theta), \hat{\alpha}^*(\theta) \rangle \, d\theta, \tag{12}$$

where $\hat{\alpha}(\theta)$ is the DFT of the discrete dislocation density $\alpha(l)$ and $\hat{\Gamma}(\theta)$ is the DFT of the discrete dislocation density $\Gamma(l)$.

3 Discrete dislocation dynamics

In this section, we apply the discrete dislocation model presented in the foregoing to the evaluation of kink nucleation energies and kink mobility on $\frac{1}{2}a\langle 111 \rangle$ screw dislocations in BCC Ta, W and Mo crystals. Dislocation kink mobility on screw segments is known to be the key rate-limiting mechanism for dislocation motion in BCC crystals. The nucleation and motion of kinks is thermally activated and, therefore, primarily controlled by the energy barriers corresponding to transitions between states. All calculations are based on force constants obtained from (Dai et al. 2006) extension of the Finnis–Sinclair potential (Finnis and Sinclair 1984). Expressions for the force constants corresponding to Embedded-Atom-Method potentials has been given in Ramasubramaniam et al. (2007), where detailed validation comparisons of dislocation core energies and structures may also be found.

3.1 Computation of incremental energies

Consider a sequence of dislocation densities α_k at discrete times t_k, $k = 0, 1, \ldots$ resulting from Monte Carlo dynamics. The evolution of the dislocation density is then governed by the energy differences $\Delta E = E_{k+1} - E_k$ between two possible successive configurations. In addition, the condition for quasistatic configurational equilibrium of a discrete dislocation density is that $\Delta E \geq 0$ for all elementary flips (Ramasubramaniam et al. 2007). Incremental energies due to flips are, therefore, of particular interest for purposes of characterizing discrete dislocation equilibrium configurations and evolution.

Starting from (12), a straightforward calculation gives

$$\begin{aligned} \Delta E &= E_{k+1} - E_k \\ &= \frac{1}{(2\pi)^3} \int_{[-\pi,\pi]^3} \frac{1}{2} \langle \hat{\Gamma}(\theta)\hat{\alpha}_{k+1}(\theta), \hat{\alpha}^*_{k+1}(\theta) \rangle d\theta \\ &\quad - \frac{1}{(2\pi)^3} \int_{[-\pi,\pi]^3} \frac{1}{2} \langle \hat{\Gamma}(\theta)\hat{\alpha}_k(\theta), \hat{\alpha}^*_k(\theta) \rangle d\theta \\ &= \frac{1}{(2\pi)^3} \int_{[-\pi,\pi]^3} \langle \hat{\Gamma}(\theta)\hat{\alpha}_k(\theta), \Delta\hat{\alpha}^*(\theta) \rangle d\theta \\ &\quad + \frac{1}{(2\pi)^3} \int_{[-\pi,\pi]^3} \frac{1}{2} \langle \hat{\Gamma}(\theta)\Delta\hat{\alpha}(\theta), \Delta\hat{\alpha}^*(\theta) \rangle d\theta \end{aligned} \tag{13}$$

where we write

$$\Delta\alpha = \alpha_{k+1} - \alpha_k \tag{14}$$

for the active loopon defining the transition under consideration. Thus, the incremental energy may be expressed in the form

$$\Delta E = E_{\text{int}} + E_{\text{loopon}} \tag{15}$$

where the first term, E_{int} represents the interaction energy between the initial dislocation distribution α_k and the active loopon $\Delta\alpha$ and E_{loopon} is the formation energy of $\Delta\alpha$.

3.2 Case I

Double kink formation and subsequent spreading constitutes a fundamental rate-limiting mechanism of plastic deformation in BCC crystals. The process starts with the nucleation of a double kink on a screw dislocation segment. Once nucleated, the kink separates within its slip planes by 'zipping' along the dislocation segment, in the process furthering the plastic deformation of the crystal. The geometry of the nucleation of a discrete double kink by the addition of an elementary loopon

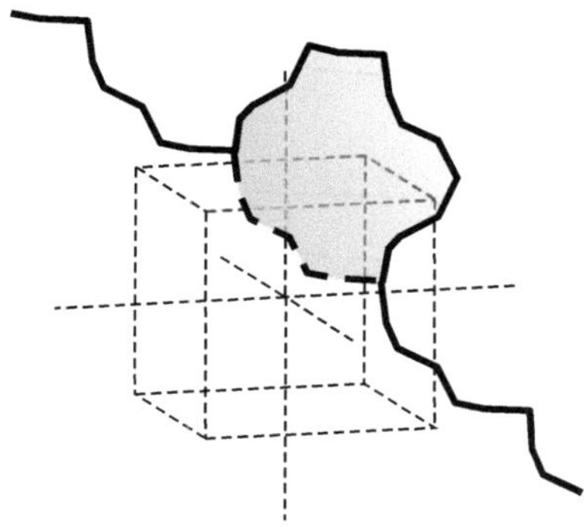

Fig. 11 Nucleation of a discrete double kink by the addition of an elementary loopon to an infinite screw discrete dislocation in a BCC lattice.

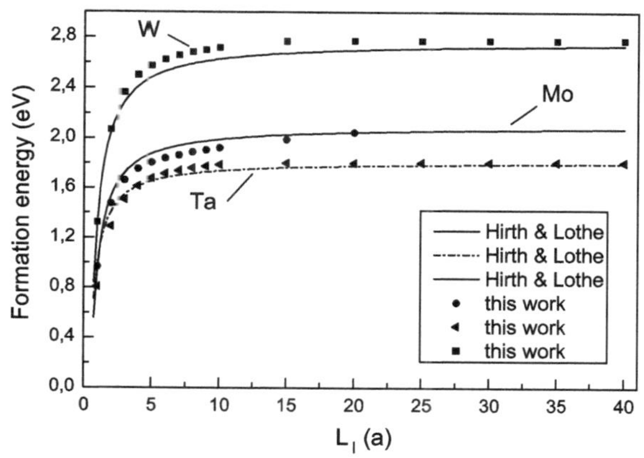

Fig. 12 Double-kink formation energy as a function of kink separation in BCC Mo, Ta and W

to an infinite screw discrete dislocation is shown in Fig. 11.

Consider, for definiteness, activity in the slip system C5 (Table 1). Consider, in addition an infinite screw dislocation defined by the sequence of 2-cells $\{e_2(4, l),\ e_2(8, l),\ e_2(10, l)\}$. The corresponding Fourier representation is

$$\begin{aligned} \hat{\alpha}(8, \theta) &= 2\pi\delta(\theta)e^{i\theta_2} \\ \hat{\alpha}(4, \theta) &= -2\pi\delta(\theta)e^{-i\theta_1} \\ \hat{\alpha}(10, \theta) &= 2\pi\delta(\theta) \end{aligned} \tag{16}$$

whereas $\Delta\hat{\alpha}(\theta)$ for the active loopon is

$$\begin{aligned} \hat{\alpha}(3, \theta) &= e^{-i(\theta_1+\theta_3)} - e^{-i(\theta_1+2\theta_3)} \\ \hat{\alpha}(4, \theta) &= e^{-i(2\theta_1+\theta_2+2\theta_3)} - e^{-i(\theta_1+\theta_3)} \\ \hat{\alpha}(5, \theta) &= e^{-i\theta_1} - e^{-i(\theta_1+\theta_3)} \\ \hat{\alpha}(8, \theta) &= e^{-i(\theta_1+\theta_3)} - e^{i\theta_2} \\ \hat{\alpha}(10, \theta) &= e^{-i(\theta_1+\theta_2+\theta_3)} - e^{-i\theta_3} \end{aligned} \tag{17}$$

The incremental energy due to the addition of a double-kink to a straight dislocation, or kink-pair formation energy, is shown in Fig. 12 as a function of kink separation. Following Hirth and Lothe (1968), the double-kink energy may be expressed as

$$E_{\text{tot}} = 2E_{\text{for}} + E_{\text{int}} \tag{18}$$

where E_{for} is the energy of formation associated with an isolated kink, and E_{int} represents the interaction energy between the kinks. By recourse to atomistic calculations, Yang et al. (2001) have obtained kink-formation energies ranging from 0.67 to 1.84 eV for 16 possible different kink configurations in tantalum. By way of comparison, the computed discrete-dislocation kink-formation energy in tantalum is 0.9 eV, which is in the ballpark of the atomistic simulations of Yang et al. (2001). Similarly, using empirical many-body potentials (Xu and Moriarty 1998) calculated the kink-pair formation energy in molybdenum to be 2 eV. Again, the corresponding discrete dislocation kink-pair nucleation energy is in good agreement with that value (Fig. 12).

As is evident from Fig. 12, the dependence of the computed double-kink formation energy on kink separation agrees remarkably well with the prediction of elasticity theory (Hirth and Lothe 1968). In particular, the interaction energy is attractive but decays in inverse proportion to the kink separation L_I and, therefore, it is negligible for well-separated kinks. It follows from these decay properties that the mobility of the individual kinks may be expected to be high once the double-kink width becomes sufficiently large, e.g., of the order of ten lattice constants. This finding is in agreement with previous atomistic calculations using an embedded atom method empirical potential (Cai et al. 2001).

3.3 Case II

Next, we wish to ascertain the nucleation energy of an elementary double-kink on a screw dislocation containing a pre-existing elementary double kink (Fig. 13). In particular, we wish to determine the separation of the incipient double kinks at which their interaction energy is negligible and, consequently, the thermally activated nucleation of the double kinks are statistically independent events.

The nucleation energy of the second double kink is shown in Fig. 14 as a function of the distance to the

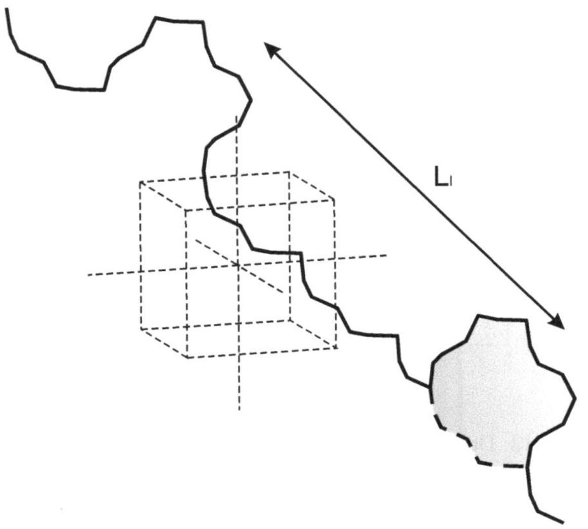

Fig. 13 Geometry of nucleation of a second elementary double kink in an infinite screw dislocation containing a pre-existing elementary double kink

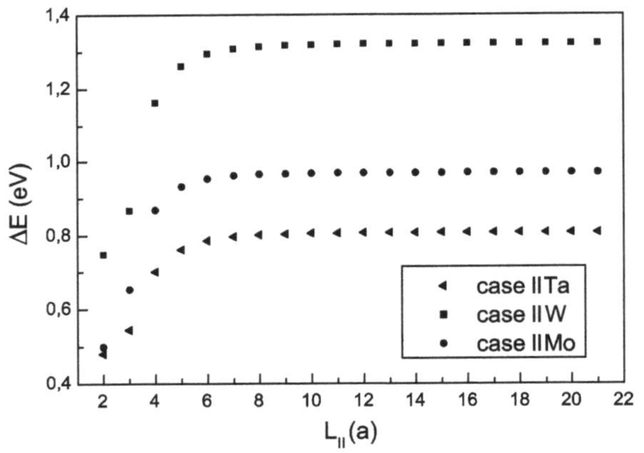

Fig. 14 Second double-kink nucleation energy as a function of distance to pre-existing elementary double kink in BCC Mo, Ta and W

pre-existing elementary kink. It is seen from the figure that the interaction between the two double kinks is attractive and decays rapidly with double-kink separation, becoming ostensibly negligible at distances greater than ten lattice distances. Interestingly, the attractive nature of the interaction energy facilitates the nucleation of double kinks in close proximity to each other, which may in turn promote clustering of double kinks on screw segments.

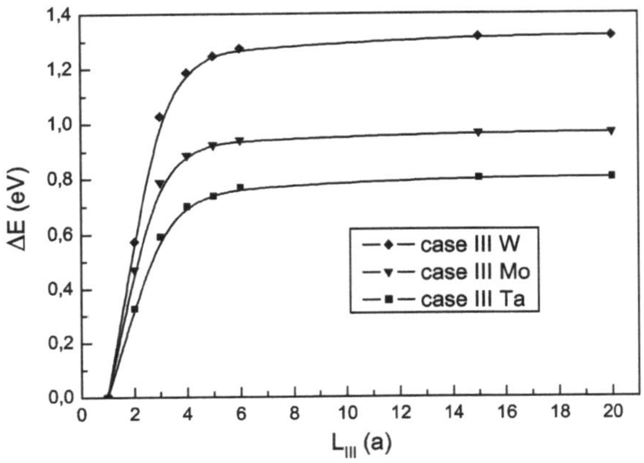

Fig. 15 Nucleation energy of the double kink as a function of the distance to the pre-existing kink

3.4 Case III

Similarly to case II, next we wish to ascertain the nucleation energy of an elementary double-kink on a screw dislocation containing a pre-existing kink. In particular, we wish to determine the distance between the pre-existing kink and the incipient kink at which their interaction energy is negligible and, consequently, the thermally activated nucleation of the kink is not influenced by the preexisting kink.

The nucleation energy of the double kink is shown in Fig. 15 as a function of the distance to the pre-existing kink. As in Case II, it is evident from the figure that the interaction between the double kink and the pre-existing kink is attractive and decays rapidly with separation, becoming ostensibly negligible at distances greater than ten lattice distances. As in Case II, it is also remarkable that the attractive nature of the interaction energy facilitates the nucleation of double kinks in close proximity to pre-existing kinks, which may in turn promote clustering of 'daughter' double kinks ahead of 'mother' kinks.

4 Summary and conclusions

We present an application of the discrete-dislocation theory of Ariza and Ortiz to the characterization of the energetics of kinks in Mo, Ta and W BCC crystals. For definiteness, we have restricted our attention to three kink and double-kink geometries. In all cases, the transitions are effected by the addition of elementary loops, termed loopons by Shenoy et al. (1999). The discrete dislocation calculations supply detailed predictions of formation and interaction energies for various double-kink formation and spreading mechanisms as a function of the geometry of the double kinks, including: the dependence of the formation energy of a double kink on its width; the energy of formation of a double kink on a screw dislocation containing a pre-existing double kink; and energy of formation of a double kink on a screw dislocation containing a pre-existing single kink. Within a thermally-activated framework, the attractive nature of the computed interaction energies between kinks and double kinks is expected to facilitates the nucleation of double kinks in close proximity to each other and to pre-existing kinks, thus promoting clustering of double kinks on screw segments and of 'daughter' double kinks ahead of 'mother' kinks.

The predictions of the discrete dislocation theory are found to be in good agreement with the full atomistic calculations based on empirical interatomic potentials available in the literature, including: the calculations of Yang et al. (2001) of kink-formation energies ranging from 0.67 to 1.84 eV for 16 possible different kink configurations in tantalum; the calculations of (Xu and Moriarty 1998) of the kink-pair formation energy in molybdenum; and previous atomistic calculations using an embedded atom method empirical potential (Cai et al. 2001). The great advantage of the discrete dislocation theory over direct atomistic simulations is that the former leads to analytical forms of the interaction energies of lattice defects that are explicit up to simple quadratures, thus conferring discrete dislocation theory a considerable computational advantage. In addition, discrete dislocation theory overcomes many of the essential difficulties that have traditionally afflicted linear-elastic dislocation theory. Thus, discrete dislocation theory may be regarded as an effective compromise between linear-elastic dislocation theory and full atomistic simulations based on empirical potentials.

Acknowledgments We gratefully acknowledge the support of the Ministerio de Ciencia e Innovación of Spain (DPI2009-14305-C02-01/02) and the support of the Consejería de Innovación of Junta de Andalucía (P09-TEP-4493). Support for this study was also provided by the Department of Energy National Nuclear Security Administration under Award Number DE-FC52-08NA28613 through Caltech's ASC/PSAAP Center for the Predictive Modeling and Simulation of High Energy Density Dynamic Response of Materials.

A The discrete Fourier transform

A.1 Definition and fundamental properties

The DFT of $f : \mathbb{Z}^n \to \mathbb{R}$ is

$$\hat{f}(\theta) = \sum_{l \in \mathbb{Z}^n} f(l) e^{-i\theta \cdot l} \tag{19}$$

In addition we have

$$f(l) = \frac{1}{(2\pi)^n} \int_{[-\pi,\pi]^n} \hat{f}(\theta) e^{-i\theta \cdot l} d^n\theta \tag{20}$$

which is the inversion formula for the DFT. It follows from this expression that $f(-l)$ is the Fourier-series coefficient of $\hat{f}(\theta)$. We have the identity

$$\widehat{f * g} = \hat{f}\hat{g} \tag{21}$$

which is often referred to as the *convolution theorem*. Suppose in addition that f, g are square-summable. Then

$$\sum_{l \in \mathbb{Z}^n} f(l) g^*(l) = \frac{1}{(2\pi)^n} \int_{[-\pi,\pi]^n} \hat{f}(\theta) \hat{g}^*(\theta) d^n\theta \tag{22}$$

which is the Parserval identity for the DFT.

References

Ariza MP, Ortiz M (2005) Discrete crystal elasticity and discrete dislocations in crystals. Arch Ration Mech Anal 178:149–226

Ariza MP, Ortiz M (2010) Discrete dislocations in graphene. J Mech Phys Solids 58:710–734

Babuska I, Vitasek E, Kroupa F (1960) Some applications of the discrete fourier transform to problems of crystal lattice deformation, parts i and ii. Czechoslovak J Phys 10:419–427; 488–504

Cai W, Bulatov V, Justo JF, Argon AS, Yip S (2000) Intrinsic mobility of a dissociated Dislocation in silicon. Phys Rev Lett 84(15):3346–3349

Cai W, Bulatov V, Yip S, Argon AS (2001) Kinetic Monte Carlo modeling of dislocation motion in BCC metals. Mater Sci Eng A 309(SI):270–273

Dai X, Kong Y, Li J, Liu B (2006) Extended Finnis–Sinclair potential for bcc and fcc metals and alloys. J Phys Condens Matter 18:4527–4542

Duesbery M (1983) On kinked screw dislocations in the bcc lattice. I. Kink energies and double kinks. Acta metallurgica 31(10):1759–1770

Ericksen JL (1979) On the symmetry of deformable crystrals. Arch Ration Mech Anal 72:1–13

Finnis M, Sinclair J (1984) A simple empirical n-body potential for transition-metals. Philos Mag A Phys Condens Matter Struct Defects Mech Prop 50(1):45–55

Ghoniem NM, Tong SH, Sun LZ (2000) Parametric dislocation dynamics: a thermodynamics-based approach to investigations of mesoscopic plastic deformation. Phys Rev B 61(2):913–927

Hirth JP, Lothe J (1968) Theory of dislocations. McGraw-Hill, New York

Koslowski M, Cuitiño A, Ortiz M (2002) A phase-field theory of dislocation dynamics, strain hardening and hysteresis in ductile single crystals. J Mech Phys Solids 50(12):2597–2635

Madec R, Devincre B, Kubin LP (2002) Simulation of dislocation patterns in multislip. Scripta Materialia 47(10):689–695

Munkres JR (1984) Elements of algebraic topology. Perseus Publishing, Cambridde, MA

Mura T (1987) Micromechanics of defects in solids. Kluwer, Boston

Nye JF (1953) Some geometrical relations in dislocated crystals. Acta Metallurgica 1:153–162

Ramasubramaniam A, Ariza MP, Ortiz M (2007) A discrete mechanics approach to dislocation dynamics in bcc crystals. J Mech Phys Solids 55(3):615–647

Seeger A (1981) The kink picture of dislocation mobility and dislocation-point-defect interactions. Journal de Physique 42(NC5):201–228

Shenoy VB, Ortiz M, Phillips R (1999) The atomistic structure and energy of nascent dislocation loops. Model Simul Mater Sci Eng 7(4):603–619

Tang M, Kubin LP, Canova GR (1998) Dislocation mobility and the mechanical response of BCC single crystals: a mesoscopic approach. Acta Materialia 46(9):3221–3235

Vitek V (1976) Computer-simulation of screw dislocation-motion in bcc metals under effect of external shear and uniaxial stresses. Proc R Soc Lond Ser A 352(1668):109–124

Wang YU, Jin YM, Cuitino AM, Khatachuryan AG (2001) Nanoscale phase field microelasticity theory of dislocations: model and 3d simulations. Acta Materialia 49(10):1847–1857

Wang YU, Jin YM, Cuitino AM, Khatachuryan AG (2004) Phase field microelasticity theory and modeling of multiple dislocation dynamics. Appl Phys Lett 78(16):2324–2326

Xu W, Moriarty J (1998) Accurate atomistic simulations of the Peierls barrier and kink-pair formation energy for ⟨111⟩ screw dislocations in bcc Mo. Comput Mater Sci 9:348–356

Yang L, Soderlind P, Moriarty J (2001) Accurate atomistic simulation of a/2 ⟨111⟩ screw dislocations and other defects in bcc tantalum. Philosoph Mag A 1(5):1355–1385

Yang LH, Moriarty JA (2001) Kink-pair mechanisms for a/2 ⟨111⟩ screw dislocation motion in bcc tantalum. Mater Sci Eng A 319(SI):124–129

Zbib HM, de la Rubia TD, Bulatov VV (2002) A multiscale model of plasticity based on discrete dislocation dynamics. Int J Mech Sci 124(1):78–87

Int J Fract (2012) 174:41–48
DOI 10.1007/s10704-011-9673-z

ORIGINAL PAPER

The expanding spherical inhomogeneity with transformation strain

Xanthippi Markenscoff

Received: 22 August 2011 / Accepted: 20 December 2011 / Published online: 17 January 2012

Abstract Within the context of linear elastodynamics, the radiated fields (including inertia) for a spherical *inhomogeneous* (of different elastic constants) inclusion with dilatational transformation strain (or eigenstrain), expanding in general motion under applied loading, can been obtained on the basis of Eshelby's equivalent inclusion method by using the strain field of the expanding *homogeneous* spherical inclusion (as a function of the eigenstrain) to determine the equivalent eigenstrain. With the equivalent dynamic eigenstrain (which is dependent on the boundary motion), the radiated fields for the inhomogeneous spherical expanding inclusion can be obtained. Based on them, the "driving force" (self-force) on the moving boundary can be computed, and this is the rate of mechanical work (with inertia) required to create an incremental region of inhomogeneity with eigenstrain, i.e with the elastic constants changing as the region of the eigenstrain expands. The self-force depends on the history of the motion, and, in the presence of external loading the driving force yields a Peach-Koehler type force, which exhibits coupling of the applied loading to the history of the motion of the boundary velocity.

Keywords Transformation strain · Inhomogeneity · Driving force · Moving phase boundary

X. Markenscoff (✉)
University of California, San Diego, La Jolla, CA 92093-0411, USA
e-mail: xmarkens@ucsd.edu

1 Introduction

The analytic solution to the elastodynamic problem of a spherical dynamically expanding in general subsonic motion Eshelby inclusion with dilatational transformation strain, i.e. a spherically expanding boundary of strain discontinuity, in an isotropic material was obtained by Markenscoff and Ni (2010) on the basis of a formula given by Willis (1965) in terms of the dynamic Green's function. Here we expand this solution to the case of an expanding spherical inhomogeneity (i.e. the material constants are different) with dilatational transformation strain (see Fig. 1). This solution is unique and has as initial condition the static (Eshelby 1957) solution, which is a minimum energy solution. The Hadamard jump conditions are satisfied on the moving surface of discontinuity inclusion.

Based on these radiated fields, the "driving force", obtained from the mechanical energy-release rate which is the self-force on the moving boundary, was calculated by Markenscoff and Ni (2010). It has two parts: the static one, coinciding with Eshelby (1970, 1977) and Gavazza (1977), and a dynamic one due to inertia, that depends on the velocity of the boundary. One may note, that the "driving force" expression of Atkinson and Eshelby (1968), Freund (1972) (for moving cracks) and Eshelby (1970) coincides for a purely mechanical system in an isothermal process with the expression for this force obtained by the phase transformation community (Abeyaratne 1983; Truskinovsky 1982; Heidug and Lehner 1985; Abeyaratne

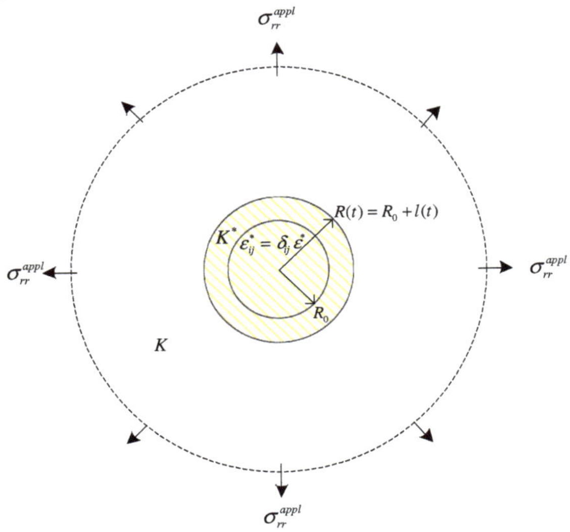

Fig. 1 The moving spherical inhomogeneity boundary with dilatational transformation strain

and Knowles 1990) for a moving boundary of discontinuity). One may note that the "contour independent" dynamic J integral (based on Noether's theorem), in the limit as it shrinks on the moving defect, is equivalent to the energy-release rate and the "driving force" (Freund 1990; Stolz 2003).

With the external loading included in the total dynamic J integral, an additional term $\langle\sigma_{k\ell}^{appl}\rangle[[\varepsilon_{k\ell}^{*}(\mathbf{x},t)]]$ is obtained in the driving force which is the counterpart of the Peach-Koehler force on dislocations (Markenscoff and Ni 2011; Markenscoff 2010a,b). Thus, the driving forces on moving defects, dislocations (Eshelby 1953) and inclusion boundaries, have exactly the same structure, a self-force due to inertia, and a Peach-Koehler force associated with the external loading. The self-force constitutes the mechanical rate of work needed to create an incremental region of eigenstrain by whatever source.

Gupta and Markenscoff (2008, 2011), showed that the variation of the energy functional to be equal to the dynamic J integral is a *necessary and sufficient* condition for linear momentum to be preserved in the domain during an incremental translation of the defect boundary. Thus, the dynamic J integral is not an *ad hoc* assumption, but a necessary and sufficient condition for linear momentum to be preserved in the domain under translation of the defect boundary. The relation between boundary velocity and external loading is governed by the vanishing of the total (including the applied loading) dynamic J integral according to Noether's theorem. In nonlinear elasticity, the total J integral is still zero according to Noether's theorem (in the absence of dissipation), but, since the applied stress field cannot be superposed as in linear elasticity, the effect of the applied loading in the driving force cannot be separated, and the relation between boundary velocity and loading will be implicit. The "Eshelby principle" (Eshelby 1970) that applying an external stress field (that renders the total driving force, or, equivalently, the total J integral, equal to zero) can create mechanically an incremental region of eigenstrain has been demonstrated by Markenscoff (2010b) by a simple example.

In this paper, the analysis of Markenscoff and Ni (2010) is extended to the expanding spherical constrained *inhomogeneous* (different elastic constants) inclusion by using the Eshelby method (1957,1961) of equivalent eigenstrain extended now to dynamics, as was done for the plane inhomogeneous boundary by Markenscoff (2011). The equivalent eigenstrain is determined by the point-wise equivalence of the stresses, expressed in terms of strains by the constitutive relation. While for ellipsoidal inclusions, the relation of the total constrained strain to the equivalent eigenstrain is through the Eshelby tensor (leading to a system of algebraic equations), in dynamics the total constrained strain is related to the equivalent eigenstrain by the dynamic solution of Markenscoff and Ni (2010, 2011). The invertibility of this relationship for dynamics may need further investigation. Knowing the equivalent eigenstrain, the elastodynamic fields of a plane inhomogeneous inclusion moving boundary with transformation strain can be calculated from the Willis (1965) formula, and, with these radiated fields, the driving forces can be obtained. Since the equivalent eigenstrain depends on the motion $l(t)$ of the boundary and the applied stress, the Peach-Koehler type force associated with the external loading now involves *coupled* terms of the loading with the motion and nonlinear terms in the applied stress.

2 Radiated fields from an expanding spherical inclusion with dilatational eigenstrain

We present here briefly the formulation and the computed fundamental solution of Markenscoff and Ni (2010) for an expanding spherical constrained homogeneous inclusion boundary with dilatational transformation strain, or equivalently called eigenstrain.

It follows the analysis of Willis (1965) treating constrained inclusions with time dependent eigenstrain. Willis (1965) did not consider inclusions with time changing domains.

In an infinite isotropic elastic body, consider a spherical inclusion centered at the origin of a three dimensional Cartesian coordinates system, and expanding in the radial direction from the radius $R(t) = R_0$ for $t \leq 0$ to the radius $R(t) = R_0 + l(t)$ for $t \geq 0$, where $l(t) > 0$ is a smooth function such that $l(0) = 0$. Let the spherical inclusion with radius $R(t)$ have a uniform dilatational distribution of eigenstrain within the sphere $V(t)$

$$\varepsilon^*_{ij}(x,t) = \begin{cases} \delta_{ij}\varepsilon^*, x \in V(t) \equiv \{x \| x| \leq R(t)\} \\ 0, \text{otherwise} \end{cases} \tag{1}$$

for all t, where $i, j = 1, 2, 3$, δ_{ij} is Kronecker delta, and ε^* is a constant.

Markenscoff and Ni (2010) obtained the fields of the expanding inclusion following the analysis of Willis (1965) treating constrained inclusions with time dependent eigenstrain, and, more specifically, Eq. (26) of Willis (1965) for the displacement field $u_i(\mathbf{x}, t)$ due to domain and time dependent eigenstrain ε^*_{ij}:

$$\begin{aligned} u_i(\mathbf{x},t) &= \int_{-\infty}^{+\infty} dt' \int_{V(t)} C_{jklm}\varepsilon^*_{lm}(\mathbf{x}';t')\frac{\partial G_{ij}(\mathbf{x}-\mathbf{x}',t-t')}{\partial x_k}dV' \\ &= \int_{-\infty}^{+\infty} dt' \int_{\partial V_t} C_{jklm}\varepsilon^*_{lm}(\mathbf{x}';t')n_k(\mathbf{x})G_{ij}(\mathbf{x}-\mathbf{x}',t-t')dS \end{aligned} \tag{2}$$

where $i, j, k, l, m = 1, 2, 3$, C_{jklm} are the elastic constants, $n_k = x_k/R(t)$ is the outward normal of the boundary ∂V_t, and G_{ij} are components of the elastic dynamic Green's function for an infinite isotropic body

$$\begin{aligned} G_{km}(\bar{r}_i,\bar{t}) = \frac{1}{4\pi\rho}\Bigg\{ &\frac{\bar{t}}{\bar{r}^2}\left[\frac{3\bar{r}_k\bar{r}_m}{\bar{r}^3} - \frac{\delta_{km}}{\bar{r}}\right] H\left(\bar{t}-\frac{\bar{r}}{c_1}\right)H\left(\frac{\bar{r}}{c_2}-\bar{t}\right) \\ &+\frac{\bar{r}_k\bar{r}_m}{\bar{r}^3}\left[\frac{\delta\left(\bar{t}-\frac{\bar{r}}{c_1}\right)}{c_1^2} - \frac{\delta\left(\bar{t}-\frac{\bar{r}}{c_2}\right)}{c_2^2}\right] \\ &+\frac{\delta_{km}\delta\left(\bar{t}-\frac{\bar{r}}{c_2}\right)}{c_2^2}\Bigg\} \end{aligned} \tag{3}$$

$$\bar{t} = t - t', \bar{r}_i = x_i - x'_i, \bar{r}^2 = (x_1 - x'_1)^2 + (x_2 - x'_2)^2 + (x_3 - x'_3)^2, \tag{4}$$

$$c_1 = \sqrt{(\lambda + 2\mu)/\rho}, \quad c_2 = \sqrt{\mu/\rho} \tag{5}$$

where $H(\cdot)$ denotes the Heaviside step function and $\delta(\cdot)$ the delta function.

We will apply Eq. (1) to a constrained inclusion with dilatational eigenstrain occupying the sphere $x_1 \leq R_0$ for $t \leq 0$ and expanding according to $R(t) = R_0 + l(t)$, such that $l(t) = 0$ for $t \leq 0$, i.e.,

$$\varepsilon^*_{lm}(\mathbf{x},t) = \varepsilon^*_{lm}H(R_0 + l(t) - r) = \delta_{lm} \times \varepsilon^* H(R_0 + l(t) - r) \tag{6}$$

where for an isotropic material the elastic coefficient tensor is

$$C_{jklm} = \lambda\delta_{jk}\delta_{lm} + \mu(\delta_{jl}\delta_{km} + \delta_{jm}\delta_{kl}). \tag{7}$$

The evaluation of the surface integrals in (2) performed by Markenscoff and Ni (Abeyaratne 1983) yielded the solution for the displacement (with a denoting the dilatational wave speed $c_1 = a$):

$$\begin{aligned} u_r(r,t) &= \frac{a(3\lambda+2\mu)\varepsilon^*}{4(\lambda+2\mu)r^2}\int_t^{\frac{r+R_0}{a}}(r^2 + R_0^2 - a^2s^2)ds \\ &\times\frac{R_0^3(3\lambda+2\mu)\varepsilon^*}{3r(\lambda_2\mu)} + \frac{a(3\lambda+2\mu)\varepsilon^*}{4(\lambda+2\mu)r^2}\int_0^{\tau_2}\left[r^2+R^2(t')-a^2(t-t')^2\right]dt' \\ &= \frac{(3\lambda+2\mu)\varepsilon^*}{4\rho a^2r^2}\Bigg\{a\left(r^2+R_0^2\right)(\tau_2 - t) \\ &+\frac{1}{3}\left[2\left(r^3+R_0^3\right)+a^3(t-\tau_2)^3\right] \\ &+\int_0^{\tau_2} a\left[2R_0l(t')+l^2(t')\right]dt'\Bigg\} \end{aligned} \tag{8}$$

where τ_1, and τ_2, $0 \leq \tau \leq t$, are respectively the unique (for subsonic motion) solutions of

$$r + R(\tau_1) = a(t - \tau_1) \tag{9}$$

$$a(t - \tau_2) = |r - R(\tau_2)| \tag{10}$$

corresponding to the times that a wavelet emitted from the back or front boundary of the expanding inclusion, respectively, has the time to reach the field point (r, t) (see Fig. 2)

The above solution recovers the static spherical Eshelby inclusion (1957) when evaluated from $t = -\infty$ to $t = 0$, and satisfies the Hadamard jumps conditions on the moving boundary as shown by Markenscoff and Ni (2010).

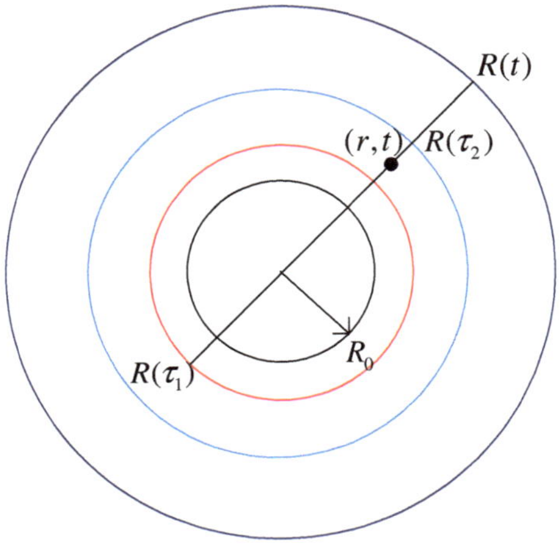

Fig. 2 Expanding spherical inclusion and contributions to the field point (r, t)

3 Radiated fields and driving forces for an expanding boundary of a spherical inhomogeneity

The radiated fields obtained by Markenscoff and Ni (2010), as given by Eqs. (8)–(10), can be used in order to obtain the radiated fields from the inhomogeneous (of different elastic constants C^*_{ijkl}) moving boundary. Eshelby (1957) equivalent inclusion method will be extended here to dynamics for a moving boundary of an expanding spherical inhomogeneity with transformation strain.

In statics, Eshelby (1957) solved the inhomogeneity problem by reducing it to an inclusion problem (same elastic constants) with equivalent eigenstrain. Let the matrix have elastic constants C_{ijkl} and the inhomogeneity C^*_{ijkl}, and consider that the inhomogeneity has eigenstrain ε^*_{ij} and is under an applied loading at infinity σ^{appl}_{ij} with corresponding strains ε^{appl}_{ij} (satisfying $\sigma^{appl}_{ij} = C_{ijkl}\varepsilon^{appl}_{kl}$). Then the total equivalent eigenstrain ε^{**}_{ij} is due to the inhomogeneity $\varepsilon^{*inhom}_{ij}$ (equivalent inclusion) plus the ("plastic") eigenstrain ε^*_{ij} in the inclusion [i.e., Mura (Markenscoff and Ni 1984), eqn. 22.12a]

$$\varepsilon^{**}_{ij} = \varepsilon^{*inhom}_{ij} + \varepsilon^*_{ij} \tag{11}$$

The Eshelby equation that determines the eigenstrain ε^{**}_{ij} of the equivalent inclusion expresses the point-wise equivalence in the stresses (of the problem with the inhomogeneity and the equivalent to it with inclusion), and is [e.g., Mura (Markenscoff and Ni 1984), eqn. 22.13a]:

$$C_{ijmn}\varepsilon^{**}_{mn} - \Delta C_{ijmn}\varepsilon_{mn}(\varepsilon^{**}_{kl}) = C^*_{ijmn}\varepsilon^*_{mn} + \sigma^{appl}_{ij} - C^*_{ijmn}\varepsilon^{appl}_{mn}, \tag{12}$$

with

$$\Delta C_{ijkl} = C_{ijkl} - C^*_{ijkl} \quad . \tag{13}$$

Equation (12), which expresses the equivalence in the stresses between inhomogeneity and inclusion, is valid for in statics and dynamics, if the total strain in Eq. (12) has the dynamic value that relates it to the eigenstrain. In the static case of the inclusion being an ellipsoid, the total strain ε_{mn} in Eq. (12) is related to the equivalent eigenstrain by the Eshelby tensor

$$\varepsilon_{mn} = S_{mnpq}\varepsilon^{**}_{pq}, \tag{14}$$

and, inserted into Eq. (12), gives a system of linear algebraic equations to be solved for the equivalent eigenstrain ε^{**}_{ij}.

In case of an inclusion with dialatational eigenstrain of Eq. (1), the equivalent eigenstrain is determined from (Mura, Eq. 22.17b)

$$K^*\left(\varepsilon^{appl}_{kk} + \varepsilon_{kk} - \varepsilon^*_{kk}\right) = K\left(\varepsilon^{appl}_{kk} + \varepsilon_{kk} - \varepsilon^{**}_{kk}\right) \tag{15}$$

where K and K^* denote the bulk moduli of the inclusion and the inhomogeneity respectively; the relation to the material constants being

$$K = \lambda + 2\mu/3 = E/3(1 - 2\nu) \tag{16}$$

$$E = 2(1 + \nu)\mu \tag{17}$$

$$\lambda = 2\mu\upsilon/(1 - 2\nu) \tag{18}$$

Equation (15) expresses the point-wise stress-strain relationship (Hooke's law), and is valid both in statics and dynamics, now with $\varepsilon_{kk}(r, t)$

$$K^*\left(\varepsilon^{appl}_{kk} + \varepsilon_{kk}(\mathbf{x}, t; l(\tau_i), \varepsilon^{**}_{kk}) - \varepsilon^*_{kk}\right) = K\left(\varepsilon^{appl}_{kk} + \varepsilon_{kk} - \varepsilon^{**}_{kk}\right) \tag{19}$$

with the total constrained strain being $\varepsilon_{mn}(\mathbf{x}, t; l(\tau_i), \varepsilon^{**}_{kk})$, where τ_i is the solution of the wavelets that had the time to reach the field point (r, t) and to contribute to the strain at that point, according to Eqs. (9) and (10).

In the case of a spherically expanding inclusion according to $R(t) = R_0 + l(t)$ such that $l(t) = 0$ for $t \leq 0$, the relation between the constrained strain and the eigenstrain is found by obtaining the strains from the displacement following Markenscoff and Ni (2010).

By differentiating the displacement the strain $\varepsilon_{rr}(r,t)$ is obtained

$$\begin{aligned}\varepsilon_{rr} &= \frac{\partial u_r}{\partial r} \\ &= -\frac{2R_0^3(3\lambda+2\mu)\varepsilon^*}{3r^3(\lambda+2\mu)}H(r-R_0-at) \\ &+ \frac{(3\lambda+2\mu)\varepsilon^*}{3(\lambda+2\mu)}H(R_0-r-at) \\ &+ \frac{\partial u_r^{II}}{\partial r}H(at-|r-R_0|)H(r+R_0-at) \\ &+ \frac{\partial u_r^{III}}{\partial r}H(at-r-R_0)\end{aligned} \tag{20}$$

$$\begin{aligned}\frac{\partial u_r^{II}(r,t)}{\partial r} &= \frac{(3\lambda-2\mu)\varepsilon^*}{2r^3(\lambda+2\mu)} \\ &\times\left[\frac{1}{3}\left(r^3+3atR_0^2-2R_0^3-a^3(t-\tau_2)^3\right)\right. \\ &\left.-\int_0^{\tau_2} aR^2(t')dt'\right]+\frac{a(3\lambda+2\mu)\varepsilon^*}{4r^2(\lambda+2\mu)} \\ &\times\left[\frac{r^2+R^2(\tau_2)-a^2(t-\tau_2)^2}{\dot{l}(\tau_2)\mp a}\right]\end{aligned} \tag{21}$$

$$\begin{aligned}\frac{\partial u_r^{III}(r,t)}{\partial r} &= \frac{(3\lambda+2\mu)\varepsilon^*}{2r^3(\lambda+2\mu)} \\ &\times\left[\frac{a^3}{3}\left((t-\tau_1)^3-(t-\tau_2)^3\right)\right. \\ &\left.-\int_{\tau_1}^{\tau_2} aR^2(t')dt'\right]+\frac{a(3\lambda+2\mu)\varepsilon^*}{4r^2(\lambda+2\mu)} \\ &\times\left[\frac{r^2+R^2(\tau_2)-a^2(t-\tau_2)^2}{\dot{l}(\tau_2)\mp a}\right. \\ &\left.+\frac{r^2+R^2(\tau_1)-a^2(t-\tau_1)^2}{\dot{l}(\tau_1)+a}\right]\end{aligned} \tag{22}$$

for $\pm(r-R(\tau_2))>0$ (equivalently, $\pm(r-R(t))>0$ corresponding to $\pm$ above.

The strains $\varepsilon_{\theta\theta}$ and $\varepsilon_{\varphi\varphi}$ are obtained from Eq. (8)

$$\varepsilon_{\theta\theta}=\varepsilon_{\varphi\varphi}=\frac{u_r}{r} \tag{23}$$

The equivalent eigenstrain can determined from Eq. (19) provided that it is invertible. This may be dependent on the motion and requires further investigation and proof. With the equivalent eigenstrain, the radiated fields for the moving boundary of an inhomogeneity with eigenstrain can be evaluated from Eq. (2):

$$u_i(\mathbf{x},t)=\int_{-\infty}^{+\infty}dt'\int_{V(t)}C_{jklm}\varepsilon_{lm}^{**}(\mathbf{x}',t';l(t'))\times\frac{\partial G_{ij}(\mathbf{x}-\mathbf{x}',t-t')}{\partial x_k}dV', \tag{24}$$

where $V(t)$ is the spherical inhomogeneity with boundary moving according to $R(t)=R_0+l(t)$.

Expression (12) should enter into Eq. (24), according to which the radiated fields of an expanding inhomogeneity boundary in general subsonic motion $l(t)$ can be evaluated, most probably numerically.

With those fields, that depend on the motion $l(t)$ in a more complex way than in the case of the inclusion boundary (Markenscoff and Ni 2010, 2011), the driving force per unit area of the inhomogeneity boundary, expressing the mechanical rate of work (per unit area) required to create an incremental volume of inhomogeneity with eigenstrain, will be computed from the mechanical energy-release rate on a contour surrounding the moving boundary and shrinking into it. For a moving discontinuity boundary, the energy-release-rate acquires the expression (Abeyaratne 1983) (also, see, Fried 1998; Stolz 2003)

$$f=[[W]]-\langle\sigma_{ij}\rangle\left[\left[\frac{\partial u_i}{\partial x_j}\right]\right], \tag{25}$$

where the brackets denote jumps, and where we defined the symbol $\langle\rangle$ by

$$\langle A\rangle=\frac{1}{2}(A^+ + A^-), \tag{26}$$

and with the strain energy jump being

$$[[W]]=\frac{1}{2}[[\sigma_{ij}\varepsilon_{ij}]]=\frac{1}{2}\sigma_{ij}\left(\frac{\partial u_k}{\partial x_m}-\varepsilon_{km}^{**}\right), \tag{27}$$

with

$$\sigma_{ij}=C_{ijkm}\left(\frac{\partial u_k}{\partial x_m}-\varepsilon_{km}^{**}\right). \tag{28}$$

Based on (27) and (28), the driving force expression of Eq. (25) reduces to (Markenscoff and Ni 2011):

$$f=-\langle\sigma_{km}\rangle\left[\left[\varepsilon_{km}^{**}(\mathbf{x},t)\right]\right], \tag{29}$$

For a spherical homogeneous inclusion the driving force (self-force) was found by Markenscoff and Ni (2010) to depend on the contribution from the back boundary of the expanding inclusion when it had the time to reach the front one

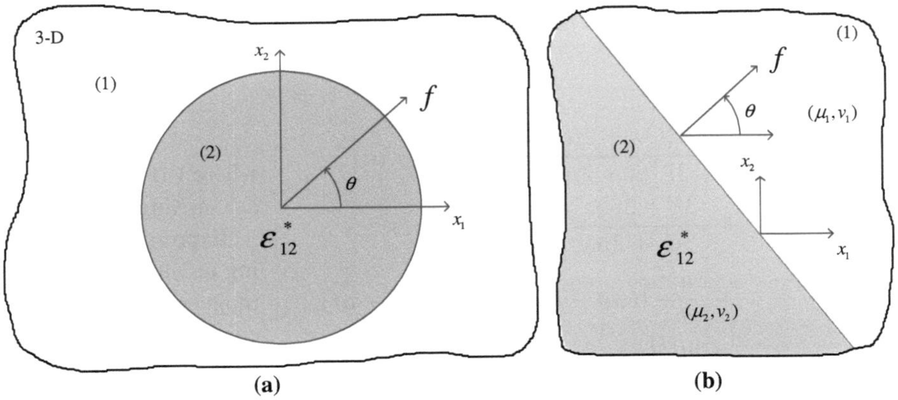

Fig. 3 Self force dependence on angle θ

$$
\begin{aligned}
f &= \varepsilon^*(3\lambda + 2\mu)\left(\left\langle\frac{\partial u_r}{\partial r}\right\rangle + 2\left\langle\frac{u_r}{r}\right\rangle\right) \\
&\quad -\frac{3}{2}(3\lambda + 2\mu)\varepsilon^{*2} \\
&= -\frac{2\mu(3\lambda + 2\mu)\varepsilon^{*2}}{(\lambda + 2\mu)} \\
&\quad -\frac{(3\lambda + 2\mu)^2\varepsilon^{*2}}{2(\lambda + 2\mu)}\left[\frac{a\dot{l}(t)}{(a^2 - \dot{l}^2(t))}\right] \\
&\quad + H(at - (R(t) + R_0))\frac{(3\lambda + 2\mu)^2\varepsilon^{*2}}{2(\lambda + 2\mu)} \\
&\quad \times \left[\frac{R(\tau_1)}{R(t)}\left(\frac{\dot{l}(\tau_1)}{a + \dot{l}(\tau_1)}\right)\right]\Big|_{r=R(t)} \qquad (30)
\end{aligned}
$$

However, in the expanding inhomogeneity case, the transformation strain is not constant, but is given by Eq. (12), which, combined with (24) implies that the self-force on the expanding inhomogeneity boundary will depend on the whole integrated history of the motion.

In the case of an applied superposed loading at infinity, the total driving force will be

$$
f^{self} + \left\langle\sigma^{appl.}_{k\ell}\right\rangle \varepsilon^{**}_{k\ell}(\mathbf{x}, t), \qquad (31)
$$

with f^{self} computed from the radiated field quantities obtained above, while, in the presence of loading due to an applied field of uniform tensile radial expansion $\sigma^{appl}_{rr} = \sigma^{appl}$, Eq. (29), which is with all the interaction energy terms included, will yield the additional term

$3\varepsilon^{**}\sigma^{appl}$

which is the counterpart of the Peach-Koehler force for dislocations.

From the structure of Eq. (57), we see that the Peach-Koehler type force term $\sigma^{appl}\varepsilon^*_{ij}$ (associated with the applied stress field) is not involving the velocity of the boundary in the case of a *homogeneous* inclusion moving boundary with constant eigenstrain ε^*_{ij}, but, in the case of the *inhomogeneously* expanding boundary, since the equivalent eigenstrain ε^{**}_{ij} in Eq. (31) depends, through Eq. (19), both on the history of the boundary motion and the applied superposed stress, there will be *coupling terms of the applied loading with the boundary velcocity* which involve the history of the motion of the boundary, as well as nonlinear terms in the applied stress.

The evolution equation of a spherical inhomogeneity under radial tensile applied loading will be

$$
\begin{aligned}
&-\frac{2\mu(3\lambda+2\mu)\varepsilon^{*2}}{(\lambda+2\mu)} - \frac{(3\lambda+2\mu)^2\varepsilon^{*2}}{2(\lambda+2\mu)}\left[\frac{a\dot{l}(t)}{(a^2-\dot{l}^2(t))}\right] \\
&+H(at - (R(t) + R_0))\frac{(3\lambda+2\mu)^2\varepsilon^{*2}}{2(\lambda+2\mu)} \\
&\times\left[\frac{R(\tau_1)}{R(t)}\left(\frac{\dot{l}(\tau_1)}{a+\dot{l}(\tau_1)}\right)\right]\Big|_{r=R(t)} \\
&+3\varepsilon^*\sigma^{appl} = \begin{cases} 0, & \text{without dissipation} \\ F(\dot{l}), & \text{dissipation or surface energy} \end{cases}
\end{aligned}
$$

Of course, in this problem spherical symmetry has been considered both on the motion and in the transformation strain, as well as in the isotropy of the inhomogeneity, so that the spherical shape is, of course, preserved.

However, for a (static) spherical inclusion with shear eigenstrain ε^*_{12} (Fig. 3), the spherical shape is not preserved, and we can calculate the self-force as a function of the angle θ, based on the Eshelby (1961) interior and exterior fields. We obtain for an inclusion

$$f(\theta) = -2\mu\left[\frac{7-5\nu}{15(1-\nu)} - \frac{1-\cos 4\theta}{8(1-\nu)}\right]\varepsilon_{12}^{*2}. \quad (32)$$

For spherical inhomogeneity of constants(μ_2, ν_2) the self-force is the same as in (32) with $\varepsilon_{12}^{*} = \varepsilon_{12}^{**}$, where

$$\varepsilon_{12}^{**} = \frac{\mu_2 \varepsilon_{12}^{*}}{2(\mu_2 - \mu_1)S_{1212} + \mu_1}, \quad (33)$$

according to Eshelby (1961), with the Eshelby tensor component $S_{1212} = \frac{4-5\nu}{15(1-\nu)}$.

For two-dimensional circular inhomogeneities with shear eigenstrain, the self-force is obtained to be

$$f(\theta) = -\frac{2\mu_2 \varepsilon_{12}^{*2}(1+\cos 4\theta)}{1+\Gamma\kappa_1} \quad (34)$$

with the stress fields computed from Dundurs and Markenscoff (2009), and where $\Gamma = \mu_2/\mu_1$, and κ, the Kolosov constant $\kappa = 3-4\nu$ in plane strain and $\kappa = (3-\nu)/(1+\nu)$ in plane stress, and where material "2" is the inhomogeneity, and material "1" the matrix.

For a plane boundary at an angle θ to the eigenstrain (Fig. 3b), the self-force is the same as the one at the point at an angle θ for a spherical/circular one. According to Eq. (34) there is an instability with $f = 0$ at $\theta = 45$ degrees.

It may be also noted that, while the elastodynamic solution for the moving inclusion boundary has been calculated analytically for an isotropic material, the solution for some classes of anisotropic materials may possibly be obtained partially analytically, as in the analysis of a generally moving dislocation in a cubic/hexagonal crystal by Markenscoff and Ni (1984, 1987), where the solution was obtained by Laplace transforms and the analysis of the Riemann surfaces corresponding to the four roots of the determinantal equation. Thus, the applicability to physical phenomena could be expanded. An application of the expanding inhomogeneous inclusion is in the modeling of earthquake sources (e.g., Burridge and Willis 1969), or in stress wave-induced phase transformations (e.g. Escobar et al. 2009)

References

Abeyaratne R (1983) An admissibility condition for equilibrium shocks in finite elasticity. J Elast 13:175–184

Abeyaratne R, Knowles JK (1990) On the driving traction acting on a surface of strain discontinuity in a continuum. J Mech Phys Solids 38:345–360

Atkinson C, Eshelby JD (1968) The flow of energy into the tip of a moving crack. Int J Fract Mech 4:3–8

Burridge R, Willis JR (1969) The self-similar problem of the expanding crack in an anisotropic solid. Proc Camb Philos Soc 66 443–468

Dundurs J, Markenscoff X (2009) Stress fields and Eshelby forces on half-plane inhomogeneities and strip inclusions meeting a free surface. Int J Solids Struct 46:2481–2485

Eshelby JD (1953) The equation of motion of a dislocation. Phys Rev 90:248–255

Eshelby JD (1961) Elastic inclusions and inhomogeneities. Prog Solid Mech 2:89–140

Eshelby JD (1957) The determination of the elastic field of an ellipsoidal inclusion and related problems. Proc R Soc Lond A 241:376–396

Eshelby JD (1970) Energy relation of the energy-momentum tensor in continuum mechanics. In: Kanninen MF et al (ed) Inelastic behavior of solid, pp 77–115

Eshelby JD (1977) Lectures on the elastic energy-momentum tensor. In: Markenscoff X, Gupta A (eds) Collected works of J.D. Eshelby: the mechanics of defects and inhomogeneities. Springer, Netherlands, pp 907–931 (2006)

Freund LB (1972) Energy-flux into the tip of an extending crack in an elastic solid. J Elast 2:341–349

Freund LB (1990) Dynamic fracture mechanics. Cambridge University Press, Cambridge

Fried E (1998) Introduction. In: Ball JM, Kinderlehler D, Podio-Guidugli P, Slemrod M (eds) Evolving phase interfaces in solids: fundamental contributions to the continuum theory of evolving phase interfaces in solids. Springer, Berlin pp 1–29

Gavazza SD (1977) Forces on pure inclusion and Somigliana dislocations. Scripta Metallurgica 11:979–981

Gupta A, Markenscoff X (2008) "Configurational forces as dissipative mechanism: a revisit" with A Gupta. Comptes Rendus Mecanique 336:126–131

Gupta A, Markenscoff X (2011) A new interpretation of configurational forces. J Elast. doi:10.1007/s10659-011-9359-9

Heidug WK, Lehner FK (1985) Thermodynamics of coherent phase transformation in nonhydrostatically stressed solids. Pure Appl Geophys 123:91–98

Hill R (1961) Discontinuity relations in mechanics of solids. In: Sneddon IN, Hill R (eds) Progress in solid mechanics. North-Holland, Amsterdam

Markenscoff X (1980) The transient motion of a nonuniformly moving dislocation. J Elast 10:193–205

Markenscoff X (1998) On the shape of the Eshelby inclusions. J Elast 49:163–166

Markenscoff X (2010a) Driving forces on phase boundaries: the Eshelby principle for an interface. Int J Fract 165:223–227

Markenscoff X (2010b) Evolution equation for moving defects: dislocations and inclusion boundaries. Int J Fract 166:35–40

Markenscoff X, Ni L (1984) Nonuniform motion of an edge dislocation in an anisotropic solid, I. Q Appl Math 41:475–494

Markenscoff X, Ni L (1987) The transient motion of a dislocation in a solid of general anisotropy. J Wave Motion 9:191–197

Markenscoff X, Ni L (2010) The energy-release rate and self-force of dynamically expanding spherical and plane inclusion boundaries with dilatational eigenstrain. J Mech Phys Solids 58:1–11

Markenscoff X, Ni L (2011) Driving forces and radiated fields for expanding/shrinking half-space and strip inclusions with general eigenstrain. Q Appl Math 69:529–548
Mura T (1982) Micromechanics of defects in solids. Martinus Nihjoff Publishers, The Hague
Stolz C (2003) Energetical approaches in nonlinear mechanics. Lectures notes, Center of Excellence for Advanced Materials and Structures. Warsaw, Poland
Truskinovsky L (1982) Equilibrium phase interfaces. Sov Phys Dokl 27:551–553
Willis JR (1965) Dislocations and inclusions. J Mech Phys Solids 13:377–395
Yang S-Y, Escobar J, Clifton RJ (2009) Computational modeling of stress wave induced martensitic transformations. Math Mech Solids 14:220–257

Int J Fract (2012) 174:49–60
DOI 10.1007/s10704-012-9693-3

ORIGINAL PAPER

A new model of damage: a moving thick layer approach

Claude Stolz · Nicolas Moës

Received: 30 August 2011 / Accepted: 16 February 2012 / Published online: 8 March 2012

Abstract A new formulation of a damage law is proposed based on a continuous transition between a sound material and a totally or partially broken material. The evolution of damage is then associated with a moving layer. This point of view permits the description of initiation and propagation of defects in an unified framework. The motion of the thick layer is defined in the frame of the moving surface Γ_o separating the sound material and the damaged material. When the damage parameters are continuous functions of the distance to Γ_o, they satisfy the conditions of transport. For particular geometries and loadings the evolution of the system is discussed. Comparison with description of damage with discontinuities and sharp interface is also presented.

C. Stolz (✉)
Laboratoire de Mécanique des Solides,
Ecole Polytechnique, CNRS UMR 7649,
91128 Palaiseau Cedex, France
e-mail: stolz@lms.polytechnique.fr

C. Stolz
LAMSID, EdF-CEA-CNRS UMR 2832, Clamart, France

N. Moës
Institut GEM, Ecole Centrale de Nantes, CNRS UMR
6183, 44321 Nantes Cedex 3, France
e-mail: nicolas.moes@ec-nantes.fr

Keywords Nucleation of defects · Moving layers · Elastic-brittle material

1 Introduction

Fracture mechanics is not sufficient generally to model the full scenario of the degradation of solids under mechanical loading. Crack initiation for instance requires damage description for modelling the gradual loss of stiffness.

When the damage evolves some regularization process are needed to avoid spurious localization. Some constitutive laws based on second gradient description of damage (Bourdin et al. 2000) or phase field approach (Karma et al. 2001; Miehe et al. 2010) have been recently proposed for this purpose. Many other approaches have been proposed to regularize the damage evolution (Pijaudier Cabot and Bazant 1987; Peerlings et al. 1996).

In this paper we propose to model the damage like the propagation of level set surface (Moës et al. 2011). For elastic quasi-brittle material, the evolution of the interface separating the undamaged material $d = 0$ from the damaged material $d = 1$ have been studied using an energetical description of the propagation (Pradeilles-Duval and Stolz 1991, 1995). In this description the parameter d jumps instantaneously from 0 to 1. The interface has no thickness and some mechanical quantities present discontinuities. Criteria of uniqueness and of stability of the propagation have been also established. In this approach, nucleation of defects can be considered as a bifurcation of equilibrium solution (Stolz 2007).

Descriptions of moving interfaces and of moving layers are also useful for studying loss of matter. Such a thermodynamical description of thin or thick layers permits to describe complex process of wear between two bodies in relative motion (Stolz 2010).

The initial material and the damaged material are separated by a surface Γ_o. This boundary is a moving interface. A surface is considered as an isopotential or a level-set. This surface is an interface where the material begins to change its properties. Two cases may be considered: the propagation of a sharp interface and the propagation of a moving layer.

- *Sharp interface*. If the transition zone is very thin, the transition from sound material to damaged material can be discontinuous and the interface is called sharp interface. The propagation of such an interface has been studied (Pradeilles-Duval and Stolz 1991, 1995).
  The thermodynamical force associated to the propagation of the interface is the local energy release rate $G(s)$. Moreover stability and non-bifurcation conditions for the problem of evolution are given when a normality rule is associated with a generalized Griffith's criterion like $G(s) \leq G_c$.
- *Moving layer*. If the interface is thick, the transition should be more regular. The evolution of damage is then associated to a moving layer. The thickness of the transition zone provides a length scale in the modelization.

The purpose of this article is to define the motion of a layer according to specific damage behaviour. The analysis of dissipation provides the definition of the energy release rate $G(s)$ associated to the motion of the layer. Other expressions of the dissipation are also given in terms of *Eshelby*'s momentum tensor. A normality rule based on a criterion function of $G(s)$ is then proposed. The evolution of damage inside a bar is then studied in this framework after comparison with sharp interface modelization.

2 A level set based damage model

We consider classical constitutive equations for describing the stress–strain relations and we propose a new formulation for the damage evolution.

The main idea of the modelization is to consider that damage occurs in a layer of finite thickness l evolving from 0 to l_c. Damage is described by a parameter d which varies from 0 to 1. The boundary between the sound material and the damaged material is denoted by Γ_o which is moving accordingly to the damage evolution. The surface Γ_o is the level-set $\phi(\mathbf{M}, t) = 0$, where $\phi(\mathbf{M}, t)$ is a signed distance function of point $\mathbf{M}$ to Γ_o. The damage d is considered as a function of the level-set ϕ.

2.1 A simple local damage model

The free energy w per unit of volume depends on the strain $\boldsymbol{\epsilon}$ and on a scalar variable d for the damage: $w = w(\boldsymbol{\epsilon}, d)$. The state law is given by

$$\boldsymbol{\sigma} = \frac{\partial w}{\partial \boldsymbol{\epsilon}}; \quad Y = -\frac{\partial w}{\partial d}, \tag{1}$$

where $\boldsymbol{\sigma}$ is the stress and Y the local energy release rate, they are functions of the state variables $(\boldsymbol{\epsilon}, d)$. For instance, we may use a potential of the form

$$w(\boldsymbol{\epsilon}, d) = \frac{1}{2}\boldsymbol{\epsilon} : \mathcal{C}(d) : \boldsymbol{\epsilon}, \tag{2}$$

then the equations of state are given as

$$\boldsymbol{\sigma} = \mathcal{C}(d) : \boldsymbol{\epsilon}; \quad Y = -\frac{1}{2}\boldsymbol{\epsilon} : \frac{\partial \mathcal{C}}{\partial d} : \boldsymbol{\epsilon}. \tag{3}$$

The thermodynamical force $Y(\boldsymbol{\epsilon}, d)$ must be finite for any $d \in [0, 1]$, this condition of regularity ensures that the dissipation is defined.

The simple form of $\mathcal{C}(d)$ is given by choosing $\mathcal{C} = d\,\mathcal{C}_1 + (1-d)\,\mathcal{C}_o$ or $\mathcal{C}^{-1} = d\,\mathcal{C}_1^{-1} + (1-d)\,\mathcal{C}_o^{-1}$. When d increases monotonically from 0 to 1, this describes the transition between a material of elastic properties $\mathcal{C}_o$ to a partially damaged material of elastic properties $\mathcal{C}_1$. For total damage description $\mathcal{C}_1 = 0$. Such behaviours are generalizations of behaviour considered in Moës et al. (2011).

The evolution of damage is determined by a dissipation potential $\mathrm{D}(\dot{d})$ convex function of $\dot{d}$ such that $\mathrm{D}(0) = 0$. The thermodynamical force Y satisfies the normality rule

$$\mathrm{D}(\dot{d}) + Y(d^* - \dot{d}) \leq \mathrm{D}(d^*), \quad \forall d^*, \tag{4}$$

which is equivalent to the relation $Y = \dfrac{\partial \mathrm{D}}{\partial \dot{d}}$ in the case of a regular potential D. These properties ensure positive dissipation. A dual formulation is given in terms of dual potential $D^*(Y) = sup_{\dot{d}}(Y\dot{d} - D(\dot{d}))$ then $\dot{d} = \frac{\partial D^*}{\partial Y}$.

For a domain of reversibility defined by $Y - Y_c \leq 0$, and a time independent constitutive relation, the normality rule is expressed as

$$\dot{d} \geq 0, \quad Y - Y_c \leq 0, \quad (Y - Y_c)\dot{d} = 0. \tag{5}$$

2.2 A new modelization

A function level set ϕ is introduced. The level set $\phi = 0$ separates the domain Ω into a undamaged and a damaged zone, this surface is denoted Γ_o.

The whole body Ω is then decomposed into three parts, the undamaged body Ω_1, the zone Ω_c of transition, the damaged body Ω_2 when damage is established. In this case, the boundary $\partial\Omega_c$ is decomposed into Γ_o where $d = 0, \phi(\mathbf{M}_o^t, t) = 0$, and Γ_1 where $d = 1, \phi(\mathbf{M}^t, t) = l_c$.

In the damaged zone Ω_c the variable d is an explicit function of the level set. The damage increases as the level set value rises:

$$\begin{cases} d(\phi) = 0, & \phi \leq 0, \quad \mathbf{X} \in \Omega_1, \\ d'(\phi) \geq 0, & 0 \leq \phi \leq l_c, \quad \mathbf{X} \in \Omega_c, \\ d(\phi) = 1, & \phi \geq l_c, \quad \mathbf{X} \in \Omega_2, \end{cases} \tag{6}$$

$d'(\phi)$ is the derivative of d relatively to ϕ. The function d is assumed to be continuous for clarity. Discontinuous function can be chosen in the proposed approach but the obtained expressions must be modified with additional contributions.

As the damage is a given continuous function of the signed distane ϕ to Γ_o, the evolution of the layer is then associated to the motion of the surface Γ_o.

2.3 The problem of equilibrium

Consider a body Ω. The external boundary $\partial\Omega$ is decomposed into two parts: $\partial\Omega_u$ where the displacement $\mathbf{u}$ is prescribed $\mathbf{u}^d(t)$ and $\partial\Omega_T$ where the loading $\mathbf{T}^d(t)$ is applied; $\partial\Omega = \partial\Omega_u \cup \partial\Omega_T, \emptyset = \partial\Omega_u \cap \partial\Omega_T$.

The total potential energy $\mathcal{E}$ is a function of $(\mathbf{u}, \phi)$

$$\mathcal{E}(\mathbf{u}, \phi) = \int_{\Omega} w(\boldsymbol{\epsilon}(\mathbf{u}), d(\phi))\, d\Omega - \int_{\partial\Omega_T} \mathbf{T}^d.\mathbf{u}\, ds, \tag{7}$$

$\boldsymbol{\epsilon}(\mathbf{u})$ is the strain, symmetric part of $\nabla\mathbf{u}$. The displacement $\mathbf{u}$ satisfies the boundary conditions on $\partial\Omega_u$: $\mathbf{u} = \mathbf{u}^d$.

Continuity conditions (CC). The displacement $\mathbf{u}$, the stress vectors $\boldsymbol{\sigma}.\mathbf{n}$, the modulus $C(d)$ are continuous functions accross the surface Γ_o. Then the gradient of the displacement and the free energy are continuous functions along each iso-ϕ surface and inside the layer. These properties imply that the velocity $\dot{\mathbf{u}}$ is also continuous. Therefore virtual variations of the displacement and of the moving surface $\phi = 0$ are uncoupled.

2.4 General motion of a layer

A point $\mathbf{M}_o^t$ on the surface Γ_o has curvilinear coordinates $s^\alpha, \alpha = 1, 2$. The local basis is defined by the tangent vectors $\mathbf{T}_\alpha = \frac{d\mathbf{M}_o^t}{ds^\alpha}$ and the normal vector $\mathbf{N} = \nabla\phi$ to the surface $\phi(\mathbf{M}_o^t, t) = 0$ (Fig. 1). A point $\mathbf{M}$ at the distance ϕ of the curve Γ_o has coordinates such that

$$\mathbf{M} = \mathbf{M}_o^t(s^\alpha) + \phi\, \mathbf{N}(s^\alpha). \tag{8}$$

In Ω_c an element of volume satisfies

$$d\Omega = j\, d\phi\, ds, \tag{9}$$

where ds is the element area at point $\mathbf{M}_o$. Due to the curvature of the surface the jacobian j is $\det(\boldsymbol{I} - \phi\bar{\boldsymbol{\kappa}})$, where $\bar{\boldsymbol{\kappa}}$ is the curvature tensor of the surface Γ at point $\mathbf{M}_o$.

In 2D, Γ_o is a curve described by the curvilinear coordinate s with tangent vector $\mathbf{T}$ and curvature κ given by

$$\frac{d\mathbf{T}}{ds} = \kappa\, \mathbf{N}. \tag{10}$$

The local derivatives of the elementary vector are

$$\begin{aligned} d\mathbf{M} &= d\mathbf{M}_o + \mathbf{N} d\phi - \phi\kappa\mathbf{T}\, ds \\ &= (1 - \kappa\phi)\,\mathbf{T}\, ds + \mathbf{N} d\phi. \end{aligned} \tag{11}$$

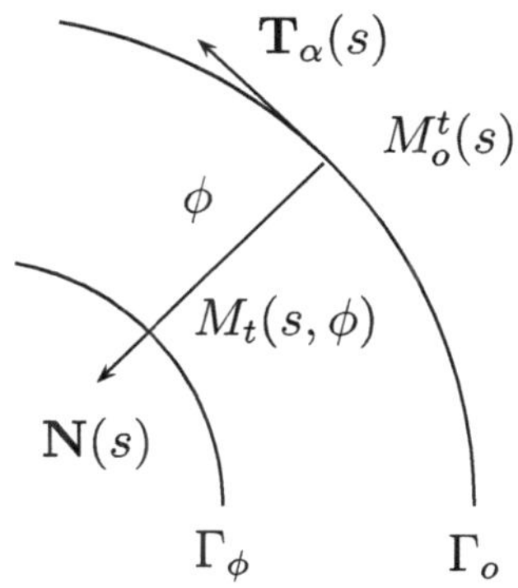

Fig. 1 Local geometry of the layer

these expressions define the local metric and the elementary area is given by $dA = (1-\phi\kappa(s))\mathrm{d}\phi\ \mathrm{d}s$ where $\kappa(s)$ is the curvature at point s of the surface $\phi = 0$.

The motion of the local geometry. When the curve Γ_o is moving, the local basis is changing accordingly. At time $t^+ = t + \Delta t$, the point $\mathbf{M}_o^t$ comes in $\mathbf{M}_o(t^+)$ such that $\phi(\mathbf{M}_o(t^+), t+\Delta t) = 0$ then $\mathbf{M}_o(t^+) = \mathbf{M}_o^t - \Delta t\ a\ (s^\alpha)\mathbf{N}$ where the normal speed a satisfies the relation

$$\frac{\partial \phi}{\partial t} - a\nabla\phi.\mathbf{N} = 0. \quad (12)$$

Any quantity f defined at a point $\mathbf{M}_o^t$ is varying during this motion. Introducing the derivative of f following the motion of Γ_o by

$$D_a f = \lim_{\Delta t \to 0} \frac{f(\mathbf{M}_o(t^+)) - f(\mathbf{M}_o^t)}{\Delta t}. \quad (13)$$

We obtain for the evolution of the geometry

$$D_a\mathbf{T} = -\frac{\mathrm{d}a}{\mathrm{d}s}\mathbf{N}, \quad D_a\mathbf{N} = \frac{\mathrm{d}a}{\mathrm{d}s}\mathbf{T},$$
$$D_a\kappa = -\frac{\mathrm{d}^2 a}{\mathrm{d}s^2} - \kappa^2 a \quad (14)$$

In an analogous way, any function $F(\mathbf{M}, t)$ defined at a point of $\mathbf{M}$ of Ω_c is variyng during this motion. Introducing the derivative of F following the motion of Γ_o by

$$D_a F = \lim_{\eta \to 0} \frac{F(\mathbf{M} + \eta D_a\mathbf{M}, t+\eta) - F(\mathbf{M}, t)}{\eta} = \frac{\partial F}{\partial t} + \frac{\partial F}{\partial \mathbf{M}}.D_a\mathbf{M}$$

where $D_a\mathbf{M} = D_a\mathbf{M}_o + \phi D_a\mathbf{N} = -a\ \mathbf{N} + \phi\ \frac{\mathrm{d}a}{\mathrm{d}s}\ \mathbf{T}$.

The position of the point $\mathbf{M}_o^t$ of Γ_o satisfies the equation $\phi(\mathbf{M}_o^t, t) = 0$, then $D_a\phi(\mathbf{M}_o^t, t) = 0$ that is exactly (12). The point $\mathbf{M}_t = \mathbf{M}_o^t + z\ \mathbf{N}$ is on the iso-level set $\phi(\mathbf{M}, t) = z$, the evolution of this surface is given by $D_a\phi(\mathbf{M}_t, t) = 0$ and obviously the normal velocity of this surface is also $a(s)$. As ϕ is a distance function, $||\mathbf{N}|| = 1$, then the variations $\delta\phi$ of ϕ are functions only of s and are denoted δa.

Variations of the potential energy. At time t, the value of the field ϕ is known. The function d is known, and $\dot{d} = \frac{\partial d}{\partial \phi}\frac{\partial \phi}{\partial t} = -d'(\phi)a(s)$. The variations of d are then $\delta d = -d'(\phi)\delta a(s)$.

The total variations of $\mathcal{E}$ with respect to the displacement and to the function ϕ taking account of the state laws are

$$\frac{\partial \mathcal{E}}{\partial \mathbf{u}}.\delta\mathbf{u} + \frac{\partial \mathcal{E}}{\partial \phi}.\delta a = \int_\Omega \boldsymbol{\sigma} : \boldsymbol{\epsilon}(\delta\mathbf{u})\,\mathrm{d}\Omega - \int_{\partial\Omega_T} \mathbf{T}.\delta\mathbf{u}\ \mathrm{d}s - \int_{\Omega_c} Y d'(\phi)\delta a\ \mathrm{d}\Omega.$$

Due to the continuity conditions (CC) the variations of displacement and the variations of the surface $\phi = 0$ are uncoupled.

Equilibrium of the system. For given ϕ, the system is that of a non-homogeneous elastic medium under boundary conditions ($\mathbf{T}^d$ along $\partial\Omega_T$ and $\mathbf{u}^d$ over $\partial\Omega_u$). The solution of the problem of equilibrium is determined by the solution of minimization of the potential energy. Then, for a state of equilibrium, the variations of the total potential with respect to $\mathbf{u}$ are zero

$$\frac{\partial \mathcal{E}}{\partial \mathbf{u}}.\delta\mathbf{u} = \int_\Omega \boldsymbol{\sigma} : \boldsymbol{\epsilon}(\delta\mathbf{u})\ \mathrm{d}\Omega - \int_{\partial\Omega_T} \mathbf{T}^d.\delta\mathbf{u}\ \mathrm{d}s = 0, \quad (15)$$

where the variations $\delta\mathbf{u}$ satisfy the boundary conditions $\delta\mathbf{u} = 0$ over $\partial\Omega_u$. By integration by parts equations of equilibrium are recovered

$$\mathrm{div}\,\boldsymbol{\sigma} = 0, \quad \mathbf{n}.\boldsymbol{\sigma} = \mathbf{T}^d, \quad \text{over } \partial\Omega_T. \quad (16)$$

Evolution of the layer. The variations of $\mathcal{E}$ are then reduced to those with respect to ϕ

$$\frac{\partial \mathcal{E}}{\partial \phi}.\delta a = -\int_{\Omega_c} Y d'(\phi)\ \delta a\ \mathrm{d}\Omega = -\int_{\Gamma_o} G(s)\delta a(s)\ \mathrm{d}s, \quad (17)$$

where the release rate of energy appears associated to the motion of the layer

$$G(s) = \int_0^l Y d'(\phi)\ j\ \mathrm{d}\phi. \quad (18)$$

In 2D situation, this expression becomes

$$G(s) = \int_0^l Y d'(\phi)\ (1 - \phi\kappa(s))\ \mathrm{d}\phi. \quad (19)$$

This defines the variation of the energy when the layer is moving. The thickness at time t is $l(t)$ less than l_c. The description is valid from the initiation of the layer from zero-thickness to a finite thickness l_c.

When l tends to 0, the thermodynamical force $G(s)$ tends to 0, and the dissipation vanishes. Definition of the evolution law must be compatible with this fact.

2.5 Dissipation of the system and evolution law

The dissipation has the classical expression

$$D_m = \int_{\partial\Omega} \mathbf{n}.\boldsymbol{\sigma}.\dot{\mathbf{u}} \text{ ds} - \frac{\mathrm{d}}{\mathrm{d}t}\int_{\Omega} w \text{ d}\Omega, \tag{20}$$

where the free energy w is a function of the strain $\boldsymbol{\epsilon}$ and of the scalar parameter d as defined (1). The internal parameters evolve only on Ω_c. Using the condition of equilibrium and the motion of the layer the total dissipation is reduced to:

$$D_m = \int_{\Omega_c} Y\dot{d} \text{ d}\Omega + \int_{\partial\Omega_c} ([w] - \boldsymbol{\sigma} : [\boldsymbol{\epsilon}])D_a\mathbf{M}.\mathbf{N} \text{ ds}. \tag{21}$$

Due to the CC, the free energy is continuous along $\partial\Omega_c$, then the last term vanishes.

The motion of the surface $\phi(\mathbf{M}_t, t) = z$ is given by the normal velocity a, and satisfies the condition $D_a\phi = 0$. The damage parameter d is a function of ϕ, then $D_a d = 0$, and the dissipation is reduced to

$$\begin{aligned} D_m &= \int_{\Omega_c} Y\nabla d.\mathbf{N}\, a \text{ d}\Omega = \int_{\Gamma_o}\int_{o}^{l} Yd'(\phi)\, a(s) j \text{ d}\phi \text{ ds} \\ &= \int_{\Gamma_o} G(s)a(s) \text{ ds}. \end{aligned} \tag{22}$$

This ensures that

$$D_m = \int_{\Gamma} G(s)a(s) \text{ ds} = -\frac{\partial\mathcal{E}}{\partial\phi}.a(s). \tag{23}$$

The derivative of the potential energy relative to ϕ is the thermodynamical force associated to the motion of the layer. This is the generalization of the expression of the thermodynamical forces defined previously for moving interface to moving layer.

On the Eshelby momentum tensor. Introducing the Eshelby momentum tensor $\boldsymbol{P}$:

$$\boldsymbol{P} = w\,\boldsymbol{I} - \boldsymbol{\sigma}.\nabla\mathbf{u}, \tag{24}$$

this tensor satisfies the relations

$$\operatorname{div}\boldsymbol{P}^T = -Y\nabla d, \text{ over } \Omega_c, \ \mathbf{N}.[\boldsymbol{P}] = 0, \text{ along } \partial\Omega_c. \tag{25}$$

Using the fact that $\nabla d = d'(\phi)\nabla\phi = d'(\phi)\mathbf{N}$ the dissipation takes the form

$$D_m = -\int_{\Omega_c} \boldsymbol{P} : \nabla D_a\mathbf{M} \text{ d}\Omega + \int_{\partial\Omega_c} \mathbf{N}.\boldsymbol{P}.D_a\mathbf{M} \text{ ds}. \tag{26}$$

These expressions can be used in numerical simulations.

Another expression of the dissipation. It can be noticed that from the expression of the local free energy:

$$d_m = \frac{1}{2}(\dot{\boldsymbol{\epsilon}} : \boldsymbol{\sigma} - \boldsymbol{\epsilon} : \dot{\boldsymbol{\sigma}}) = Y\dot{d}. \tag{27}$$

The integration of d_m over the layer is then the global dissipation

$$D_m = \int_{\Omega_c} d_m \text{ d}\Omega = \int_{\Omega_c} \frac{1}{2}(\dot{\boldsymbol{\epsilon}} : \boldsymbol{\sigma} - \boldsymbol{\epsilon} : \dot{\boldsymbol{\sigma}}) \text{ d}\Omega. \tag{28}$$

Under the hypothesis that the local damage satisfies the condition $D_a d = 0$, the dissipation takes the form

$$\begin{aligned} D_m &= \int_{\partial\Omega_c} \frac{1}{2}(\mathbf{N}.\boldsymbol{\sigma}.\nabla\mathbf{u} - \mathbf{N}.\nabla\boldsymbol{\sigma}.\mathbf{u}).D_a\mathbf{M} \text{ ds} \\ &\quad - \int_{\Omega_c} \frac{1}{2}(\boldsymbol{\sigma}.\nabla\mathbf{u} - \nabla\boldsymbol{\sigma}.\mathbf{u}) : \nabla D_a\mathbf{M} \text{ d}\Omega \end{aligned}$$

When the thickness l_c vanishes, the volume term vanishes and D_m is reduced to

$$D_m = \lim_{l_c\to 0^+} \int_{\partial\Omega_c} \frac{1}{2}(\mathbf{N}.\boldsymbol{\sigma}.\nabla\mathbf{u} - \mathbf{N}.\nabla\boldsymbol{\sigma}.\mathbf{u}) : D_a\mathbf{M} \text{ ds}.$$

Using the relation

$$\int_V \boldsymbol{\sigma} : \boldsymbol{\epsilon}\, n_i \text{ ds} = \int_V \mathbf{n}.\nabla_i\boldsymbol{\sigma}.\mathbf{u} + \mathbf{n}.\boldsymbol{\sigma}.\nabla_i\mathbf{u} \text{ ds}, \tag{29}$$

at the limit $D_a\mathbf{M} \to -a\mathbf{N}$ the dissipation becomes

$$D_m = \frac{1}{2}\int_{\Gamma_o} [\boldsymbol{\sigma}]_\Gamma : \boldsymbol{\epsilon} - \boldsymbol{\sigma} : [\boldsymbol{\epsilon}]_\Gamma \text{ ds}\, a, \tag{30}$$

that is the dissipation along a moving interface between two materials with distinct elastic properties. The model of quasi-brittle elastic material (Pradeilles-Duval and Stolz 1995) is recovered.

To define the evolution of the system, we need a link between the thermodynamical force $G(s)$ and the velocity $a(s)$.

On the evolution law. We propose now an evolution law to determine the value of velocity a. Consider the normality rule (5). From integration of $(Y - Y_c)\dot{d} = 0$ over the thickness $l \neq 0$ we get

$$\Big(\int_o^l Yd'(\phi)j\text{d}\phi - Y_c\int_o^l d'(\phi)j\text{d}\phi\Big)a = 0. \tag{31}$$

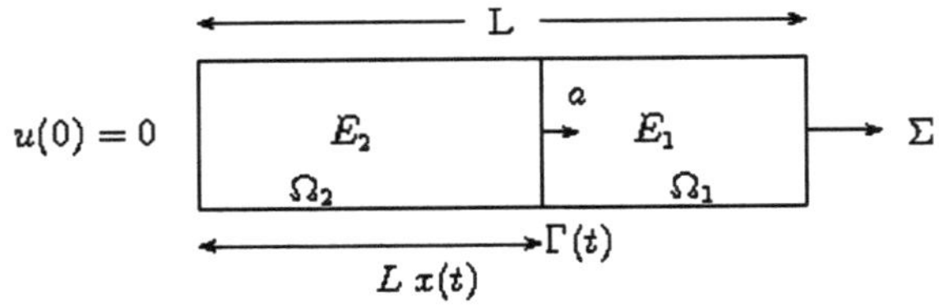

Fig. 2 A sharp interface

This provides a normality law on $G(s)$

$$(G(s) - g_c(s))a = 0, \quad a \geq 0, \quad G(s) - g_c(s) \leq 0. \tag{32}$$

For $l > 0$, $\int_o^l d'(\phi)j \, d\phi$ is positive, we can introduce the notation $\langle F \rangle$:

$$\langle F \rangle = \frac{\int_o^l F d'(\phi)j \, d\phi}{\int_o^l d'(\phi)j \, d\phi}, \tag{33}$$

with this driving force the normality rule is rewritten as

$$(\langle Y \rangle - Y_c)a = 0, \quad a \geq 0, \quad \langle Y \rangle - Y_c \leq 0. \tag{34}$$

Regularity Conditions on Y (RC) . For a regular function $Y(\epsilon, d)$, it can be noticed that when the thickness l vanishes, the value of $\langle Y \rangle$ remains finite and satisfies:

$$\lim_{l \to 0^+} \langle Y \rangle = Y(\epsilon, 0^+), \quad \lim_{l \to 0^+} \langle Y_c \rangle = Y_c \tag{35}$$

Then the critical values for initiation and for evolution of damage are the same.

The thickness l evolves from 0 to l_c. The surface $l = l_c$ is the surface defined by $d(x, t) = 1$. The transition from $d = 0$ to $d = 1$ along the normal vector $\nabla \phi$ is smooth. This approach in essentially non-local, but here the non-locality is restricted to the layer. Any classical constitutive law described by $w(\epsilon, d)$ can be used.

To study the evolution of a system under a loading history, we need two functions $C(d)$ and $d(\phi)$ or $C(\phi)$ and $d(\phi)$, because $d(\phi)$ is a continuous increasing function of ϕ.

3 Comparison of sharp and layer model

3.1 A model with sharp interface

A bar of length L is now considered. As depicted on Fig. 2, the bar is made of two linear elastic materials and is submitted to pure tension. The Young modulus of the sound material is E_1 and for the damaged material the Young modulus is E_2.

At point $X = L$, the applied tension is Σ. At point $X = 0$, displacement is $u(0) = 0$. The position of the interface is denoted Γ. We consider that a fraction $x = \Gamma/L$ of the bar is transformed.

The equilibrium and constitutive laws give:

$$\sigma = E_1\epsilon_1 = E_2\epsilon_2 = \Sigma. \tag{36}$$

And the total potential energy is

$$W(\Sigma, x) = -\frac{1}{2} L\Sigma^2 \left(\frac{x}{E_2} + \frac{1-x}{E_1} \right). \tag{37}$$

Differentiation of the energy at equilibrium with respect to the motion of interface is

$$G_\Gamma(\Sigma) = -\frac{\partial W}{\partial \Gamma} = \frac{1}{2}\Sigma^2 \left(\frac{1}{E_2} - \frac{1}{E_1} \right); \tag{38}$$

consequently the dissipation is

$$D_m = G_\Gamma(\Sigma)\dot{\Gamma}. \tag{39}$$

The speed $\dot{\Gamma} = a$ is determined by the normality law

$$a \geq 0, \quad G_\Gamma - G_c \leq 0, \quad (G_\Gamma - G_c)a = 0. \tag{40}$$

The velocity a will be positive if the consistency condition $G_\Gamma - G_c = 0$ is satisfied. This condition determines the critical value of Σ from (49):

$$\Sigma_c = \sqrt{\frac{2G_c E_1 E_2}{E_1 - E_2}}. \tag{41}$$

Initially the bar is composed of material 1 only. Assuming that initial conditions are $x(0^+) = 0$ and $u(L, 0^+) = 0$. Prescribing the displacement at point $X = L$, the total strain of the bar is $\Delta(t)$ and satisfies the global constitutive law

$$\Delta(t) = \frac{u(L,t) - u(0,t)}{L} = \Sigma \left(\frac{x}{E_2} + \frac{1-x}{E_1} \right). \tag{42}$$

The response of the bar is given by

$$\begin{cases} \Delta \leq \Delta_c = \dfrac{\Sigma_c}{E_1}, & \Sigma = E_1\Delta, \quad x = 0, \\ \dfrac{\Sigma_c}{E_2} \geq \Delta \geq \Delta_c, & \Sigma = \Sigma_c, \quad \dot{x} = \dot{\Delta}\dfrac{E_1 E_2}{\Sigma_c(E_1 - E_2)}, \\ \Delta \geq \dfrac{\Sigma_c}{E_2}, & \Sigma = E_2\Delta, \quad x = 1. \end{cases}$$

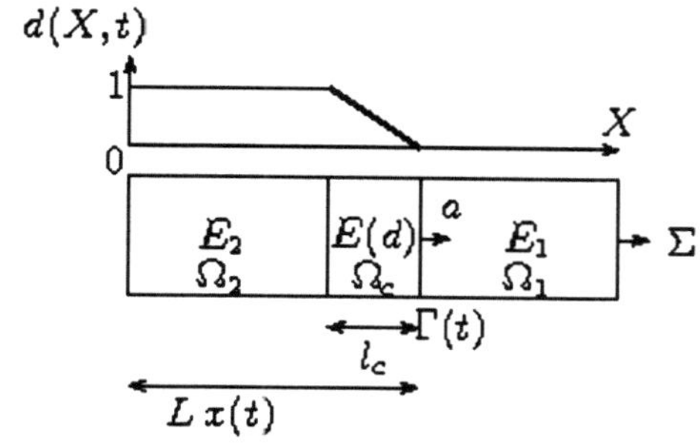

Fig. 3 The propagation of a layer

3.2 A model with a moving layer

The free energy w for unixial response is

$$w(\epsilon, d) = \frac{1}{2}E(d)\epsilon^2, \tag{43}$$

where d varies from 0 to 1, the Young modulus $E(d)$ is a continuous function of d, then there is no discontinuity at $d = 0$. On Fig. 3, $\phi(X,t) = \Gamma(t) - X$ and the value of the damage parameter is given by

$$\begin{cases} \phi(X,t) \le 0, & d(\phi) = 0, \\ 0 \le \phi(X,t) \le l_c, & d(\phi) = \phi/l_c, \\ \phi(X,t) \ge l_c, & d(\phi) = 1. \end{cases} \tag{44}$$

The damage parameter d is an increasing function of the distance ϕ to the boundary Γ separating the sound material to the damaged one. Generally, the function $d(\phi)$ is a given continuous function of ϕ. As the Young modulus E is a continuous function of d and for comparison with the sharp interface, it is natural to consider the matching conditions $E(0) = E_1$ and $E(1) = E_2$.

Initially $\Gamma(t) = 0$ and the propagation of the layer begins at the origin of the bar, so the thickness $\Gamma(t) = l(t)$ is smaller than l_c. The thickness increases to l_c and after this step of initialization, the thickness is kept constant.

For the given constitutive laws, the dissipation is local and only due to damage $d_m = Y\dot{d}$. Integrating over the layer we get the total dissipation due to damage inside the bar:

$$D_m = \int_o^l Y\dot{d}\, \mathrm{d}\phi. \tag{45}$$

Assuming that d is a continuous function of ϕ.

Using the constitutive law, the local driving force Y is

$$Y = -\frac{1}{2}\frac{dE}{dd}\epsilon^2 = -\frac{\Sigma^2}{2E^2(d)}\frac{dE}{dd}. \tag{46}$$

The function $E(d)$ must ensure that Y is finite. Then $\frac{dE}{dd}$ is finite. We adopt the notations

$$Y_o(\Sigma) = -\frac{\Sigma^2}{2E_1^2}\frac{dE}{dd}(0^+),\ Y_1(\Sigma) = -\frac{\Sigma^2}{2E_2^2}\frac{dE}{dd}(1^-) \tag{47}$$

The motion of the level-set $\phi(X,t) = 0$ is given by the velocity $\dot{\phi} = a(s)$, the total dissipation is finally expressed as

$$D_m = \frac{a}{2}\int_c^l \left(-\frac{\Sigma^2}{2E^2(d)}\right)\frac{dE}{dd}d'(\phi)\mathrm{d}\phi = G(l,\Sigma)a, \tag{48}$$

where the release rate of energy $G(l,\Sigma)$ is

$$G(l,\Sigma) = \frac{1}{2}\Sigma^2\left(\frac{1}{E(d(l))} - \frac{1}{E(0)}\right). \tag{49}$$

When $l = l_c$, we recover the expression of the sharp interface.

Using the local normality rule, the value of Y must be smaller than Y_c. This defines the critical value for initiation of damage in a point of the bar. From (35) and (46), the corresponding critical value of Σ is Σ_o such that

$$Y_o(\Sigma_o) = -\frac{1}{2}\frac{\Sigma_o^2}{E_1^2}\frac{dE}{dd}(0^+) = Y_c. \tag{50}$$

It can be noticed that the critical value Σ_o depends of the damage law and is generally different from Σ_c. To compare the two models, we assume that the dissipation of the system is the same when the layer moves with the limit thickness l_c.

3.3 The response of the bar

As d is a given function of $\phi(X,t)$, the Young modulus E (function of d) depends on position and time and is perfectly determined as a function of $\phi(X,t)$ denoted now $E(\phi)$.

$$\begin{cases} \phi(X,t) \le 0, & d(\phi) = 0,\ E(\phi) = E_1, \\ 0 \le \phi(X,t) \le l_c, & 0 \le d(\phi) \le 1,\ E(\phi), \\ l_c \le \phi(X,t), & d(\phi) = 1,\ E(\phi) = E_2. \end{cases} \tag{51}$$

On the driving force Y. Inside the layer $0 \le \phi(X,t) \le l_c$, $\frac{dE}{dd} = E'(\phi)/d'(\phi)$ then we have for the local force Y defined by (46)

$$\begin{cases} \phi(X,t) \le 0, & Y(\phi,\Sigma) = Y_o(\Sigma), \\ 0 \le \phi(X,t) \le l_c, & Y(\phi,\Sigma) = -\dfrac{\Sigma^2}{E^2(\phi)} \dfrac{E'(\phi)}{d'(\phi)}, \\ l_c \le \phi(X,t), & Y(\phi,\Sigma) = Y_1(\Sigma). \end{cases} \quad (52)$$

The function Y is a continuous function of X.

Global answer for a given $\phi(X,t)$. Due to equilibrium $\sigma = \Sigma$, and the constitutive behaviour defines the strain

$$\epsilon(X,t) = \frac{du}{dX}(X,t) = \frac{\Sigma(t)}{E(\phi)},$$

and by integration over the length of bar, the global answer of the bar is obtained as

$$\Delta = \frac{u(L,t) - u(0,t)}{L} = \frac{1}{L}\int_o^L \epsilon(X,t)dX.$$

As previously we use the notation $x = \Gamma/L$ and $x_c = l_c/L$.

State 1. Initially, the bar is elastic with modulus E_1:

$$\Delta = \Sigma/E_1, \quad Y(X) \le Y_c.$$

State 2. A damaged zone exists with a thickness l smaller than l_c, then $\Gamma = l < l_c$ or $x \le x_c$, $\phi(X,t) = \Gamma - X$:

$$\Delta = \Sigma \int_o^{\Gamma} \frac{dX}{E(\phi)L} + \frac{\Sigma}{E_1}(1-x), \ \langle Y \rangle \le Y_c.$$

State 3. The layer is established, $l_c \le \Gamma \le L$, $x_c = l_c/L$, $\phi(X,t) = \Gamma - X$:

$$\Delta = \frac{\Sigma}{E_2}(x - x_c) + \int_{\Gamma - l_c}^{\Gamma} \frac{\Sigma}{E(\phi)} \frac{dX}{L} + \frac{\Sigma}{E_1}(1-x), \langle Y \rangle \le Y_c.$$

State 4. The layer arrives at point $\Gamma = L$, $x = 1$. $d = 1$ for $X = L - l$, $l \le l_c$, $\phi(X,t) = L + l_c - l - X$:

$$\Delta = \frac{\Sigma}{E_2}(1 - l/L) + \Sigma \int_{L-l}^{L} \frac{dX}{E(\phi)L}.$$

State 5. All the bar is damaged:

$$\Delta = \frac{\Sigma}{E_2}.$$

The constitutive law (state 1–5) can be written as

$$\Delta = S(x)\,\Sigma, \quad (53)$$

where $0 \le x \le 1$. The modulus $S(x)$ evolves with the motion of the layer, then $\dot{S} = \dfrac{dS}{dx}\dot{x} = S'(x)\dot{x}$.

3.4 The problem of evolution

During state 1–3, the mechanical state $(l, \Gamma, \Delta, \Sigma)$ must be compatible with the domain of reversibility $G(l,\Sigma) \le Y_c d(l)$. For each value of $l \le l_c$ the critical value Σ_l is defined by

$$G(l, \Sigma_l) = Y_c d(l), \quad \text{or} \quad \langle Y \rangle = Y_c. \quad (54)$$

For state 4, these conditions must be reconsidered, $\Gamma = L$ and $d = 1$ for $X = L - l$; the modulus of elasticity at $X = L$ is $E(l_c - l)$, using (49) Eq. (54) becomes:

$$G(l,\Sigma) = \frac{1}{2}\Sigma^2\left(\frac{1}{E_2} - \frac{1}{E(l_c - l)}\right) = Y_c(1 - d(l_c - l)).$$

Now, we consider only the evolution during state 1–3. It can be noticed that for $\Sigma \le \Sigma_l$ the layer cannot evolve, because $G(l,\Sigma)$ is an increasing function of Σ. The condition $F(l,\Sigma) = G(l,\Sigma) - Y_c d(l) \le 0$ implies that for $F = 0$ the evolution of F satisfies

$$\dot{F} = \left(\frac{\partial G}{\partial l}\dot{l} + \frac{\partial G}{\partial \Sigma_l}\dot{\Sigma} - Y_c\frac{dd}{dl}\dot{l}\right) \le 0. \quad (55)$$

At any state for which $F = G(l,\Sigma_l) - Y_c d(l) = 0$, the evolution of the bar is determined by

$$\dot{\Delta} = S'(x)\dot{x}\Sigma_l + S(x)\dot{\Sigma}, \quad (56)$$

$$\frac{\partial G}{\partial l}\dot{l} + \frac{\partial G}{\partial \Sigma_l}\dot{\Sigma} - Y_c\frac{dd}{dl}\dot{l} \le 0, \quad (57)$$

$$\dot{l} = L\,\dot{x} \ge 0. \quad (58)$$

For increasing x, the global Young modulus (S^{-1}) is decreasing, then $S'(x) \ge 0$.

Discussion for each possible state. The evolution depends on the state.

State 1: The solution is purely elastic.

State 2: The thickness of the layer is $l = xL$ and the consistency condition gives the value of $\dot{x}$ by solving the inequalities

$$\left(\frac{\partial G}{\partial l} - Y_c\frac{dd}{dl}\right)L\dot{x} + \frac{\partial G}{\partial \Sigma}\frac{\dot{\Delta} - S'(x)\Sigma_l\dot{x}}{S} \le 0, \quad \dot{x} \ge 0.$$

A unique solution $\dot{x} > 0$ exists for $\dot{\Delta} > 0$ and $\frac{\partial G}{\partial \Sigma} \geq 0$ if

$$\left(\frac{\partial G}{\partial l} - G_c \frac{dd}{dl}\right) L - \frac{S'(x)}{S} \frac{\partial G}{\partial \Sigma} \leq 0. \quad (59)$$

State 3: The energy release rate G does not depend on x and $d(l_c) = 1$ then the unique solution is given by $\dot{\Sigma} = 0$ and

$$\dot{\Delta} = S'(x)\, \Sigma_c^o\, \dot{x}.$$

When we apply an increasing global strain $\Delta(t)$ the bar follows successively the different states.

3.5 Matching conditions between the two models for a damaged bar

After the step of initialization, $l = l_c$. The thermodynamical forces for the two descriptions, using the sharp interface or the moving layer, are identical, because $E(0) = E_1$, $E(l_c) = E_2$

$$G(l_c, \Sigma) = G_\Gamma(\Sigma). \quad (60)$$

The motion of the layer is given by the normality rule

$$G(l, \Sigma) - Y_c d(l) \leq 0, \quad a \geq 0, \quad (61)$$

$$(G(l, \Sigma) - Y_c d(l))a = 0. \quad (62)$$

When the interface is established ($l = l_c$), the equality of the two dissipations, obtained for a sharp interface ($G_c a$) and for a moving layer ($Y_c d(l) a$), provides the value of Y_c:

$$D_m = G_c a = Y_c d(l_c) a \quad \Rightarrow Y_c = G_c\,. \quad (63)$$

Using the normality rule (62), the velocity a is positive if and only if $G(l, \Sigma) = Y_c d(l)$. This defines the critical value Σ_l of Σ for which the layers evolves. For $l = 0$ the critical value of Σ is Σ_o and for l_c the critical value is Σ_c.

As G is an increasing function of Σ at fixed l, a unique solution with a layer is obtained if $\Sigma_l \leq \Sigma_o$, otherwise all points of the elastic domain (where $d = 0$) are critical with respect to initiation.

At point $\phi(X, t) = l_c$, we obtain $Y(l_c, \Sigma) = Y_1(\Sigma)$, this ensures the continuity of Y along the bar (Figs. 4, 5).

4 Particular choice of constitutive laws

We study the reponse of the bar for two particular constitutive laws defined by specific function d and E (Figs. 4, 5).

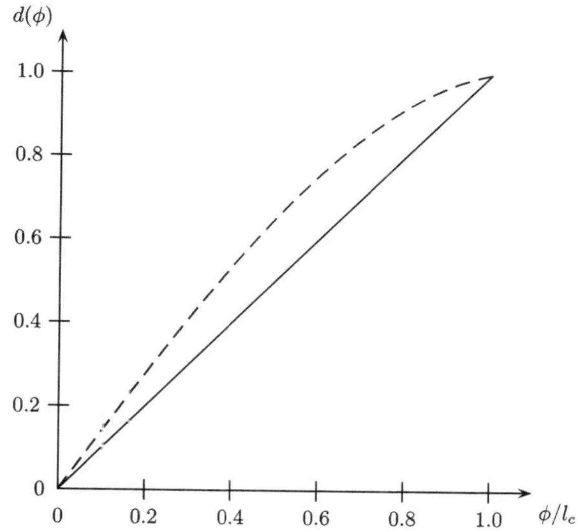

Fig. 4 Damage $d(\phi)$, $d_1 = 0$, *solid line*, $d_1 \neq 0$: *dashed line*

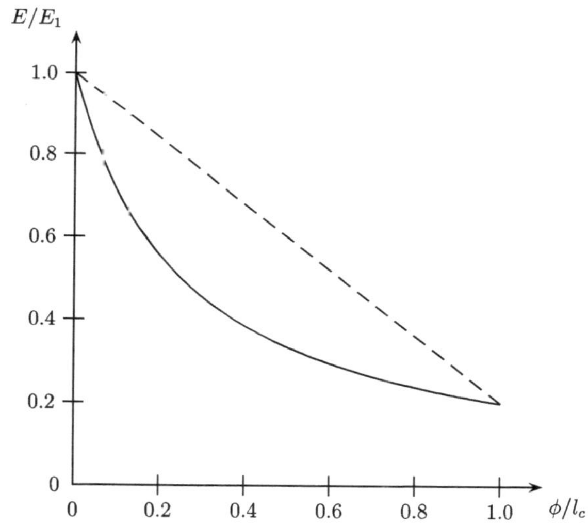

Fig. 5 Young modulus evolution $E(\phi)$, C1: *solid line*, C2: *dashed line*

– Case 1 (C1). The functions $E(\phi)$,$d(\phi)$ are given by

$$\begin{cases} \phi(X,t) \leq 0, & E(\phi) = E_1, \quad d(\phi) = 0, \\ l_c \geq \phi(X,t) \geq 0, & \begin{cases} \dfrac{1}{E(\phi)} = \dfrac{1 - \phi/l_c}{E_1} + \dfrac{\phi/l_c}{E_2} \\ d(\phi) = \dfrac{\phi}{l_c}\left(1 + d_1 - d_1 \left(\dfrac{\phi}{l_c}\right)^\beta\right) \end{cases} \\ \phi(X,t) \geq l_c, & E(\phi) = E_2, \quad d(\phi) = 1. \end{cases}$$

The definition of finite Y and $\langle Y \rangle$ implies restrictions on parameters d_1 and β: $d_1 \geq 0$ and $\beta \geq 0$, the exponent β is chosen such d' is finite $\forall \phi \in [0, l_c]$.

$$d'(\phi) = \frac{1}{l_c}\left(1 + d_1 - d_1(\beta+1)\left(\frac{\phi}{l_c}\right)^\beta\right) < \infty. \quad (64)$$

then d_1 and β satisfy: $d_1 \geq 0$ and $0 < \beta < \frac{1}{d_1}$ or $d_1 = 0$. The value Y is obtained in the layer as

$$Y(\phi, \Sigma) = -\frac{1}{2}\frac{\Sigma^2}{E^2(\phi)}\frac{E'(\phi)}{d'(\phi)}, \quad (65)$$

then $Y_o(\Sigma)$ and $Y_1(\Sigma)$ are

$$Y_o(\Sigma) = \frac{\Sigma^2}{2(1+d_1)}\frac{E_1 - E_2}{E_1 E_2},$$

$$Y_1(\Sigma) = \frac{\Sigma^2}{2(1-\beta d_1)}\frac{E_1 - E_2}{E_1 E_2}.$$

A positive velocity a is obtained if the consistency condition is satisfied

$$G(l, \Sigma_l) = Y_c d(l). \quad (66)$$

In this case, we get the relation

$$\langle Y \rangle = \frac{1}{2}\Sigma^2\left(\frac{1}{E_2} - \frac{1}{E_1}\right) = Y_c\left(1 + d_1\left(1 - \left(\frac{l}{l_c}\right)^\beta\right)\right).$$

This implies that for $d_1 = 0$, the layer initiates and after moves at $\Sigma = \Sigma_c$

$$\Sigma_c = \sqrt{\frac{2Y_c E_1 E_2}{E_1 - E_2}}. \quad (67)$$

For $d_1 > 0$

$$\Sigma_l = \Sigma_c\sqrt{\left(1 + d_1\left(1 - \left(\frac{l}{l_c}\right)^\beta\right)\right)}. \quad (68)$$

Initiation. The thermodynamical force Y for $d = 0$ is

$$Y(0, \Sigma) = Y_o(\Sigma). \quad (69)$$

At initiation the critical value Σ_o satisfies $Y_o(\Sigma_o) = Y_c$. As $E'(0) = \frac{1}{l_c}\left(\frac{1}{E_2} - \frac{1}{E_1}\right)$ and $d'(0) = (1 + d_1)/l_c$, we get

$$\Sigma_o = \Sigma_c\sqrt{1 + d_1} \geq \Sigma_c. \quad (70)$$

State 2. The local value of Y in the layer is

$$Y(\phi, \Sigma_l) = \frac{1}{2}\frac{\Sigma_l^2}{E^2(\phi)}\frac{E'(\phi)}{d'(\phi)}, \quad (71)$$

that is

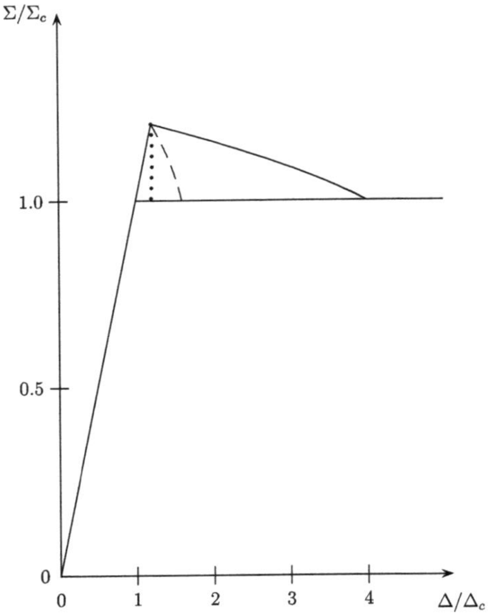

Fig. 6 Case C1, Global response, $E2/E1 = .2$, $d_1 = .45$, $\beta = 2$, $l_c = 5$, $L = 10$, *solid line*, $L = 50$, *dashed line*, $L = 250$, *dotted line*

$$Y(\phi, \Sigma_l) = \frac{1}{2}\left(\frac{1}{E_2} - \frac{1}{E_1}\right)\Sigma_l^2\frac{1}{d'(\phi)}, \quad (72)$$

after reduction, for $0 < X < l$, $\phi(X,t) = l - X$:

$$Y(\phi, \Sigma_l) = Y_c\frac{1 + d_1 - d_1\left(\frac{l}{l_c}\right)^\beta}{1 + d_1 - d_1(\beta+1)\left(\frac{\phi}{l_c}\right)^\beta}, \quad (73)$$

then along the bar Y is a decreasing function of X. It can be noticed that Y is continuous in $X = l$. Then in Ω_1 the initiation of new defect is not possible. The local value $Y(\phi, \Sigma)/Y_c$ is plotted in Fig. 7 for different value of l/l_c. This shows the evolution of the local force Y during the state 2.

On Fig. 6, the global response of the bar is plotted for $\beta = 2$, $d_1 = .45$, $l_c = 5$ and different value of ratio l_c/L, in space $(\Delta/\Delta_c, \Sigma/\Sigma_c)$, where $\Delta_c = E_1\Sigma_c$. The strain - stress curve is determined for given l:

$$\Sigma_l = \Sigma_c\sqrt{1 + d_1 - d_1\left(\frac{l}{l_c}\right)^\beta}, \quad (74)$$

$$\Delta = \Sigma_l\left(\frac{1}{E_1} + \frac{3}{2}\left(\frac{1}{E_2} - \frac{1}{E_1}\right)\left(\frac{l}{l_c}\right)\frac{l_c}{L}\right). \quad (75)$$

For this case, it is easy to show that when l_c/L is small, snap-back occurs and the condition of uniqueness (59) is not satisfied. The evolution of the local force Y during stet 2 is plotted in Fig. 7.

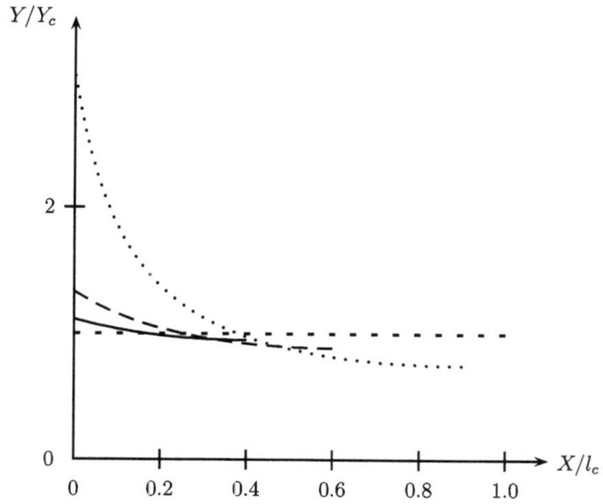

Fig. 7 Evolution of $Y(X, \Sigma_l)/Y_c$ in the layer in case C1 for $l/l_c = .3$, *solid line*; .6, *dashed line*; .9, *dotted line*

State 3. $\Sigma = \Sigma_c$, the local force $Y(\phi, \Sigma_c)$ is given by

$$\begin{cases} \phi(X,t) \leq 0, & Y(\phi, \Sigma_c) = Y_o(\Sigma_c) \leq Y_o(\Sigma_o), \\ 0 \leq \phi(X,t) \leq l_c, & Y(\phi, \Sigma_c) \\ & = \dfrac{Y_c}{1 + d_1 - d_1(\beta+1)\left(\dfrac{\phi}{l_c}\right)^{\beta}}, \\ l_c \leq \phi(X,t), & Y(\phi, \Sigma_c) = Y_1(\Sigma_c). \end{cases}$$

- Case 2 (C2). For $0 \leq \phi(X,t) \leq l_c$, consider now the expressions:

$$E(\phi) = (1 - \phi/l_c)E_1 + \phi/l_c E_2, \tag{76}$$

$$d(\phi) = \phi/l_c. \tag{77}$$

As previously, $Y(\phi, \Sigma)$ is non uniform and is a decreasing function of X, for $\phi(X,t) = l(t) - X$, with $X \leq l$:

$$Y(\phi, \Sigma) = \frac{\Sigma^2}{2}\frac{E_1 - E_2}{\left(\left(1 - \dfrac{\phi}{l_c}\right)E_1 + \dfrac{\phi}{l_c}E_2\right)^2}, \tag{78}$$

and for $d = 0$ and $d = 1$ we have:

$$Y_o(\Sigma) = \frac{1}{2}\frac{\Sigma^2}{E_1^2}(E_1 - E_2),$$

$$Y_1(\Sigma) = \frac{1}{2}\frac{\Sigma^2}{E_2^2}(E_1 - E_2).$$

The consistency condition gives the critical loading Σ_l

$$\Sigma_l^2 = \frac{2Y_c E_1}{E_1 - E_2}\left(\frac{l}{l_c}E_2 + \left(1 - \frac{l}{l_c}\right)E_1\right), \tag{79}$$

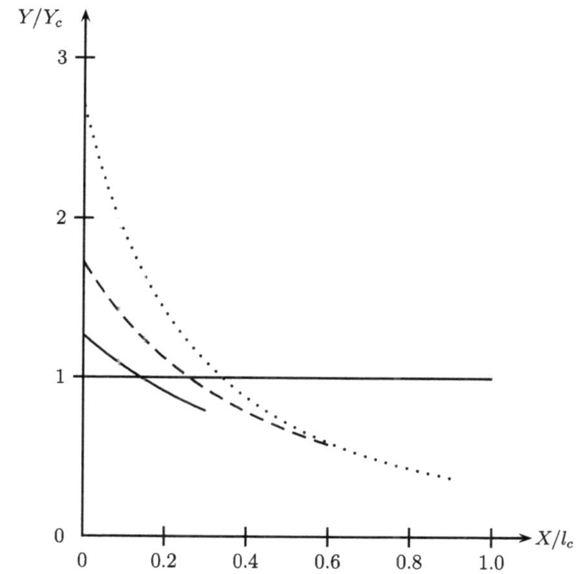

Fig. 8 Evolution of $Y(X, \Sigma_l)/Y_c$ in the layer Case C2, $E_2/E_1 = .2$, $l/l_c = .3$, *solid line*; .6, *dashed line*; .9, *dotted line*

then

$$\Sigma_0^2 = \frac{2Y_c E_1^2}{E_1 - E_2} > \Sigma_{l_c}^2 = \frac{2Y_c E_1 E_2}{E_1 - E_2} = \Sigma_c^2. \tag{80}$$

The value Σ_o of initiation of the layer is greater than the value Σ_c. When $\Sigma = \Sigma_l$

$$Y(\phi, \Sigma_l) = Y_c E_1 \frac{\left(1 - \dfrac{l}{l_c}\right)E_1 + \dfrac{l}{l_c}E_2}{\left(\left(1 - \dfrac{\phi}{l_c}\right)E_1 + \dfrac{\phi}{l_c}E_2\right)^2}, \tag{81}$$

then

$$Y(l, \Sigma_l) \geq Y_c, \quad Y(0, \Sigma_l) = Y_c. \tag{82}$$

Figure 8 shows the local force $Y(X, \Sigma_l)$ during the state 2 for different value of l/l_c along the layer. The strain–stress curve is easy to obtained:

$$\Sigma_l = \Sigma_c\sqrt{\frac{l}{l_c} + \left(1 - \frac{l}{l_c}\right)\frac{E_1}{E_2}},$$

$$\Delta = -\frac{l_c \Sigma_l}{L(E_1 - E_2)}\left(\ln\frac{l}{l_c} + \left(1 - \frac{l}{l_c}\right)\frac{E_1}{E_2}\right) + \frac{\Sigma_l}{E_1}\left(1 - \frac{l}{L}\right).$$

In this case, there is a snap-back during the growth of the layer. Typical global responses of the bar are plotted on Fig. 9 for different values of the ratio l_c/L.

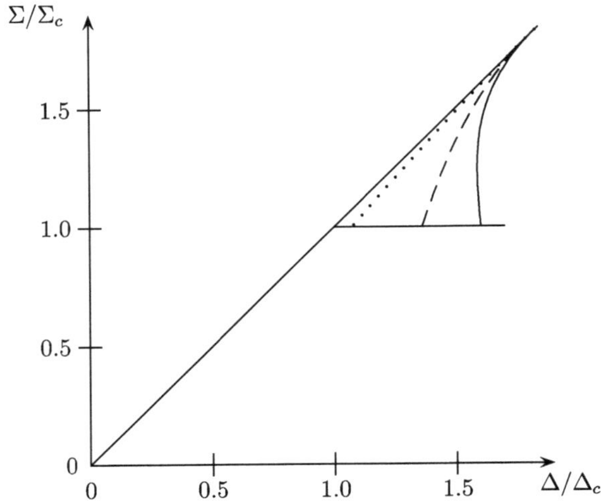

Fig. 9 Case C2. Global response. $E_2/E_1 = .2, l_c = 5, L = 10$, *solid line*, 50, *dashed line*, 250, *dotted line*

State 3. Now $\Sigma = \Sigma_c$ and Y is defined along the bar as

$$\begin{cases} \phi(X,t) \leq 0, & Y(\phi, \Sigma_c) = Y_o(\Sigma_c) \leq Y_o(\Sigma_o), \\ 0 \leq \phi(X,t) \leq l_c, & Y(\phi, \Sigma_c) \\ & \quad = \dfrac{Y_c E_1 E_2}{\left(\left(1 - \dfrac{\phi}{l_c}\right) E_1 + \dfrac{\phi}{l_c} E_2\right)^2}, \\ l_c \leq \phi(X,t), & Y(\phi, \Sigma_c) = Y_1(\Sigma_c), \end{cases}$$

where $Y_o(\Sigma_c) = Y_c E_2/E_1$ and $Y_1(\Sigma_c) = Y_c E_1/E_2$.

These two examples show the role of the constitutive law on the phase of initiation of defect. The strain–stress curve for the bar depends on the ratio E_2/E_1 and l_c/L. The solution is not necessary unique, taking account of the remark on the consistency solution.

5 Conclusion

We have studied a new approach on damage evolution based on the notion of level set and moving layer. The initiation of damage and its evolution are presented in the same framework.

Comparisons with the evolution of a sharp interface are made. The critical loading value at initiation is greater than the critical value of loading for the propagation of the moving layer. When the limit thickness l_c is reached, the critical value of the loading to obtain propagation is the same for the sharp interface and the moving layer.

Generally the dissipation and the thermodynamical forces for motion of the layer depend on the local shape of the layer, in particular on the curvature of the moving surface $d = 0$. It is kwown that the presence of curvature is a factor of stabilization of the propagation of an interface (Son et al. 1989). This will be studied in a forthcoming paper.

References

Bourdin B, Francfort G, Marigo J (2000) Numerical experiments in revisited brittle fracture. J Mech Phys Solids 48(4):797–826

Karma A, Kessler D, Levine H (2001) Phase field model of mode III dynamic fracture. Phys Rev Lett 87(4):045501

Miehe C, Welschinger F, Hofhacker (2010) Thermodynamically consistent phase field models of fracture: variational principles and multifields FE implementations. Int J Numer Methods Eng 83(10):1273–1311

Moës N, Stolz C, Bernard P, Chevaugeon N (2011) A level set based model for damage growth: the thick level set approach. Int J Numer Methods Eng 86(3):358–380

Peerlings R, de Borst R, Brekelmans W, De Vree J (1996) Gradient enhanced damage for quasi-brittle materials. Int J Numer Methods Eng 39:3391–3403

Pijaudier Cabot G, Bazant Z (1987) Non local damage theory. J Eng Mech 113:1512–1533

Pradeilles-Duval RM, Stolz C (1991) On the evolution of solids in the presence of irreversible phase transformation. Comptes Rendus Académie des Sciences de Paris, Série II 313:297–302

Pradeilles-Duval RM, Stolz C (1995) Mechanical transformation and discontinuities along a moving surface. J Mech Phys Solids 43:91–121

Son Q, Pradeilles R, Stolz C (1989) On a regularized propagation law in fracture and brittle damage. Comptes Rendus Académie des Sciences de Paris, Série II 309:1515–1520

Stolz C (2007) Bifurcation of equilibrium solutions and defects nucleation. Int J Frac 147:103–107

Stolz C (2010) Thermodynamical description of running discontinuities: application to friction and wear. Entropy 12:1418–1439

Int J Fract (2012) 174:61–74
DOI 10.1007/s10704-011-9672-0

ORIGINAL PAPER

On configurational forces at boundaries in fracture mechanics

F. D. Fischer · N. K. Simha · J. Predan · R. Schöngrundner · O. Kolednik

Received: 5 August 2011 / Accepted: 20 December 2011 / Published online: 25 January 2012

Abstract Configurational forces invariably appear at the external boundaries of cracked bodies (including the crack faces), but it is unclear whether they influence crack growth. Also, it is unclear how such boundary configurational forces are related to the J-integrals calculated in the body. In this brief note, we (i) derive expressions for the surface configurational forces and determine their values on regions of the external boundaries with prescribed tractions or displacements, (ii) determine the relation between the far-field J-integral and the surface configurational forces, and (iii) show that surface configurational forces on the crack faces do not alter the relation between the near-tip and far-field J-integrals.

Keywords Configurational forces · External boundaries · Role of crack faces · J-integral vector · Analytical solutions · Finite element modeling

F. D. Fischer (✉)
Institute of Mechanics, Montanuniversität Leoben, Franz-Josef-Strasse 18, 8700 Leoben, Austria
e-mail: mechanik@unileoben.ac.at

N. K. Simha
Department of Biomedical Engineering, University of Minnesota, 312 Church St. S.E., Minneapolis, MN 55455, USA

J. Predan
Faculty of Mechanical Engineering, University of Maribor, Smetanova 17, 2000, Maribor, Slovenia

R. Schöngrundner
Materials Center Leoben Forschung GmbH, Roseggerstrasse 12, 8700 Leoben, Austria

O. Kolednik
Erich Schmid Institute of Materials Science, Austrian Academy of Sciences, Jahnstrasse 12, 8700 Leoben, Austria

Present Address:
N. K. Simha
Medtronic Inc., 710 Medtronic Parkway, Fridley, MN, 55432-5604, USA

1 Introduction

Configurational stresses and forces are now well-established quantities in fracture mechanics, see e.g. the books and articles by Maugin (1993, 1995, 2010), Gurtin (2000) and Kienzler and Herrmann (2000). In particular, configurational forces provide a convenient framework for studying shielding (or anti-shielding) of cracks due to material inhomogeneities, eigenstrain fields, plasticity, etc., see e.g. Simha et al. (2003, 2005, 2008), Kolednik et al. (2010) and references therein.

It is well known that configurational forces invariably appear at the external boundaries of any loaded body, see the example presented in Fig. 1, a linear-elastic plate containing a hole subjected to uniform, uniaxial tension. Such surface configurational forces can be viewed as driving forces for phenomena that can cause the external boundary to move in the reference configuration resulting in changes of the shape of

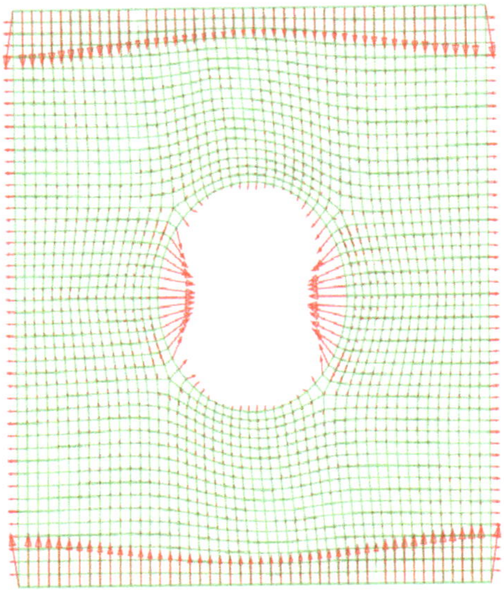

Fig. 1 Homogeneous linear-elastic plate containing a hole, subjected to uniform tension in the vertical direction. Configurational surface forces $\mathbf{f}_s$ appear at all the external boundaries. The far-field J-integral vector $\mathbf{J}_{far}$ is zero here, since the integral of $\mathbf{f}_s$ over the entire external boundary vanishes, see Eq. (22). The deformed configuration is shown, and configurational body forces appear at internal nodes, because the numerical implementation satisfies $\nabla \cdot \mathbf{S} = \mathbf{0}$ in the interior, but not $\nabla \cdot \mathbf{C} = \mathbf{0}$

the body by diffusion, or either loss of material or accumulation of new material by electrochemical processes, such as corrosion. It is clear that such processes may also influence the propagation of cracks in materials, but the subject of the current paper is a different problem: In elastic-plastic fracture mechanics, the near-tip J-integral is most often applied for characterizing the crack driving force or the intensity of the stress and strain field at the crack tip. It has been demonstrated that the J-integral is path independent for a homogeneous body with a stationary crack and traction-free crack faces, Rice (1968). This path dependence is very important for the application of the J-integral, since the evaluation of the J-integral directly at the crack tip is in most cases not possible. The magnitude of the J-integral, evaluated along a certain contour around the crack tip, and its path dependence are intimately related to the distribution of configurational forces in the body, see e.g. Simha et al. (2005, 2008). So it is not clear *a priori* whether surface configurational forces can be ignored in all fracture mechanics calculations. In particular, it is unclear whether the surface configurational forces on the crack faces influence the path dependence of the J-integral. Such surface configurational forces have been neglected in almost all applications of configurational forces for fracture mechanics so far.

The goal of this paper is to determine the role of the surface configurational forces on the external boundaries of fracture mechanics specimens. In Sect. 2, we derive the expressions for the surface configurational forces. In Sect. 3, we determine the influence of the surface configurational forces on the J-integrals in a cracked body. In Sect. 4, we focus on the crack faces and examine the role of the configurational surface forces on the crack faces. As we have previously explained the application of the configurational forces approach to fracture problems in considerable detail in previous papers by Simha et al. (2003, 2005, 2008), we merely summarize the results here.

2 The surface configurational force

As outlined in Fig. 2, we assume a region Ω that contains a bimaterial body B with a crack. The boundary of the region Ω is denoted as $\partial\Omega$. The external boundary of the body B, including the crack faces, is denoted

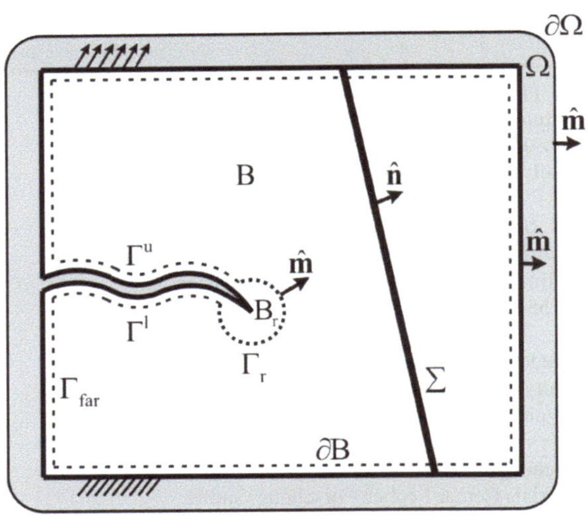

Fig. 2 The region Ω contains a bimaterial body B with a crack. The boundary of the region Ω is denoted as $\partial\Omega$. The external boundary of the body, including the crack faces, is ∂B. The bimaterial interface is Σ. The contour Γ_r encloses the crack tip at the boundary of the crack tip region B_r. The parameters $\hat{\mathbf{n}}$ and $\hat{\mathbf{m}}$ are the unit normal vectors to the interface Σ, or to the contours and boundaries, respectively. Matter of different type fill the gap between ∂B and $\partial\Omega$. The integration contours Γ^u and Γ^l along the crack faces and along the outer contour Γ_{far} of the body B are marked

as ∂B, which consists of the outer contour Γ_{far} and the parts Γ^{u} and Γ^{l} of the upper and lower crack faces. The bimaterial interface is Σ. The contour Γ_{r} with the radius r encloses the crack tip, forming the crack tip region $\mathrm{B_r}$. Note that $\mathrm{B}\backslash\mathrm{B_r}$ tends to B (excluding the crack tip), if Γ_{r} shrinks to zero. The unit vector $\hat{\mathbf{n}}$ denotes the normal to the interface Σ, while $\hat{\mathbf{m}}$ denotes the outward unit normal vector to the external boundary $\partial\mathrm{B}$, to the contour Γ_{r} of $\mathrm{B_r}$ and to the exterior boundary $\partial\Omega$. Matter of different type fills the gap between $\partial\mathrm{B}$ and $\partial\Omega$. No relevant singularity appears at the crack mouth, since the corners are considered as rounded.

The balance of deformational forces for a region Ω that contains a cracked body B without body forces (Fig. 2) requires

$$\int\limits_{\partial\Omega} \mathbf{S}\hat{\mathbf{m}}\, dl = \mathbf{0}, \tag{1}$$

where $\mathbf{S}$ denotes the first Piola-Kirchhoff stress tensor. We ignore heat conduction and inertia for simplicity. A modified version of the divergence theorem that accounts for possible discontinuities at bimaterial interfaces, external boundaries and singular stress fields at the crack tip yields

$$\int\limits_{\partial\Omega} \mathbf{S}\hat{\mathbf{m}}\, dl = \int\limits_{\Omega} \nabla\cdot\mathbf{S}\, dA + \int\limits_{\Sigma} [\![\mathbf{S}]\!]\,\hat{\mathbf{n}}\, dl + \int\limits_{\partial\mathrm{B}} [\![\mathbf{S}]\!]\,\hat{\mathbf{m}} dl + \lim_{r\to 0}\int\limits_{\Gamma_{\mathrm{r}}} \mathbf{S}\hat{\mathbf{m}}\, dl, \tag{2}$$

where ∇ is the Lagrangian gradient operator with respect to the coordinate vector $\mathbf{X}$ in the reference system and $[\![\mathbf{S}]\!]$ denotes the jump in $\mathbf{S}$ either at the bimaterial interface Σ or at the external boundary $\partial\mathbf{B}$, see the Appendices in Simha et al. (2003, 2005). Following the usual method for obtaining local relations in continuum mechanics, Eqs. (1) and (2) give

$$\nabla\cdot\mathbf{S} = \mathbf{0} \quad \text{at each point in B or } \Omega, \tag{3}$$

$$[\![\mathbf{S}]\!]\,\hat{\mathbf{n}} = \mathbf{0} \quad \text{at each point in } \Sigma, \tag{4}$$

$$[\![\mathbf{S}]\!]\,\hat{\mathbf{m}} = \mathbf{0} \quad \text{at each point in } \partial\mathrm{B}, \tag{5}$$

$$\lim_{r\to 0}\int\limits_{\Gamma_{\mathrm{r}}} \mathbf{S}\hat{\mathbf{m}}\, dl = \mathbf{0} \quad \text{at the crack tip.} \tag{6}$$

Equation (3) is the standard equilibrium equation of continuum mechanics; traction continuity on the interface Σ is ensured by Eq. (4) and on the external boundary $\partial\mathrm{B}$ by Eqs. (5), and (6) imposes restrictions on the nature of the singular stress field at the crack tip.

The total configurational force on a region Ω that contains a body B with a crack (Fig. 2) is

$$\int\limits_{\Omega} \mathbf{f}\, dA + \int\limits_{\Sigma} \mathbf{f}_{\Sigma} dl + \int\limits_{\partial\mathrm{B}} \mathbf{f}_{\mathrm{S}} dl + \mathbf{f}_{\mathrm{T}} + \int\limits_{\partial\Omega} \mathbf{C}\hat{\mathbf{m}}\, dl = \mathbf{0}, \tag{7}$$

where the configurational stress $\mathbf{C}$ acts only at bulk points in Ω, but configurational forces are presumed to act at points in the bulk ($\mathbf{f}$), at the interface ($\mathbf{f}_{\Sigma}$), the external surface including the crack flanks ($\mathbf{f}_{\mathrm{S}}$) and at the crack tip ($\mathbf{f}_{\mathrm{T}}$). The bulk configurational stress $\mathbf{C}$ is nothing but the Eshelby- or energy-momentum tensor (see e.g. the derivation in Simha et al. 2003),

$$\mathbf{C} = \phi\mathbf{I} - \mathbf{F}^{\mathrm{T}}\mathbf{S}, \tag{8}$$

where ϕ is the Helmholtz potential or the stored energy density, $\mathbf{I}$ is the identity tensor and $\mathbf{F}$ denotes the deformation gradient tensor ($\mathbf{F}^{\mathrm{T}}$ is the transposed of $\mathbf{F}$).

Applying the modified version of the divergence theorem, Eq. (2) results with $\nabla|_{\mathrm{expl}}$ being the divergence operator with respect to an explicit dependence on $\mathbf{X}$ (i.e. keeping the fields $\mathbf{F}$ and $\mathbf{S}$ fixed, see e.g. Maugin 2010) in the following:

$$\nabla|_{\mathrm{expl}}\cdot\mathbf{C} + \mathbf{f} = \mathbf{0} \quad \text{at each point in B or } \Omega, \tag{9}$$

$$[\![\mathbf{C}]\!]\,\hat{\mathbf{n}} + \mathbf{f}_{\Sigma} = \mathbf{0} \quad \text{at each point on } \Sigma, \tag{10}$$

$$[\![\mathbf{C}]\!]\,\hat{\mathbf{m}} + \mathbf{f}_{\mathrm{S}} = \mathbf{0} \quad \text{at each point in } \partial\mathrm{B}, \tag{11}$$

$$\lim_{r\to 0}\int\limits_{\Gamma_{\mathrm{r}}} \mathbf{C}\hat{\mathbf{m}}\, dl + \mathbf{f}_{\mathrm{T}} = \mathbf{0} \quad \text{at the crack tip.} \tag{12}$$

Equations (9), (10) and (12) agree with our previous derivations in Simha et al. (2003, 2005, 2008) and with other derivations in the literature, see Maugin (1993, 1995, 2010), Gurtin and Podio-Guidugli (1996), Gurtin (2000), Kienzler and Herrmann (2000), Steinmann (2000). In previous papers we did not focus on the entire external boundary of the cracked body, but examined only smooth crack faces with both faces considered as a single interface in the body. Also, we did not consider surface configurational forces along the crack faces explicitly as in Eq. (11) and instead assumed $[\![\mathbf{C}]\!]\,\hat{\mathbf{m}} = \mathbf{0}$ at the interface between the regions below and above the crack. As we will see below, for most situations the surface configurational force $\mathbf{f}_{\mathrm{S}} = \mathbf{0}$, if the crack faces are smooth; then $[\![\mathbf{C}]\!]\,\hat{\mathbf{m}} = \mathbf{0}$ is consistent with Eq. (11). However, this needs not to be so necessarily. Therefore, in the current paper we explicitly consider surface configurational forces for the entire external boundary.

Typically the matter in the region outside the body (i.e. in $\Omega\backslash\mathrm{B}$) will consist of one or more of the following: air or vacuum, rigid body, incompressible fluid, or compressible fluid. We now examine the configurational stress $\mathbf{C}$ and configurational body force $\mathbf{f}$ in the region outside the body (i.e. in $\Omega\backslash\mathrm{B}$) for the different cases:

(a) Vacuum or air: Ignoring the negligible ambient pressure in air, we have $\phi = 0$ and $\mathbf{S} = \mathbf{0}$, hence we obtain $\mathbf{C} \equiv \mathbf{0}$ and consequently $\mathbf{f} = \mathbf{0}$.
(b) Rigid body: This would correspond to the grips of a loading device that impose displacement controlled boundary conditions. Here the stored energy density $\phi = 0$ and deformation gradient $\mathbf{F} = \mathbf{I}$, but the stress $\mathbf{S} \neq \mathbf{0}$; however, the stress is not related to the stored energy density and is instead determined from the solution of the equilibrium problem. We obtain $\mathbf{C} \equiv -\mathbf{S}$ and consequently $\mathbf{f} = \mathbf{0}$.
(c) Incompressible fluid: An example would be hydraulic fracturing. This is identical to the rigid body case.
(d) Compressible fluid: Here $\mathbf{C}$ would primarily be a function of density. If the fluid is homogeneous, then $\mathbf{f} = \mathbf{0}$. However, one has to be careful, if the fluid is inhomogeneous, since the configurational body force would not vanish identically then.

Consequently, for almost all situations that are encountered in structural fracture mechanics (meaning excluding the compressible fluid case), both the bulk configurational stress and the bulk configurational force in the material outside the fracture specimen vanish,

$$\mathbf{C} = \mathbf{0} \quad \text{and} \quad \mathbf{f} = \mathbf{0} \qquad \text{in } \Omega\backslash\mathrm{B}. \tag{13}$$

Next we examine the surface configurational force $\mathbf{f}_\mathrm{S}$ on the external boundaries of the body $\partial\mathrm{B}$, which includes the crack faces. Eq. (11) can be re-written as

$$\mathbf{f}_\mathrm{S} = -[\![\mathbf{C}]\!]\,\hat{\mathbf{m}} = -\left(\mathbf{C}^{\mathrm{out}} - \mathbf{C}^{\mathrm{in}}\right)\hat{\mathbf{m}} \\ = \left(\mathbf{C}^{\mathrm{in}} - \mathbf{C}^{\mathrm{out}}\right)\hat{\mathbf{m}}. \tag{14}$$

$\mathbf{C}^{\mathrm{in}}$ is the limiting value of $\mathbf{C}$, when $\partial\mathrm{B}$ is approached from inside the body B. $\mathbf{C}^{\mathrm{out}}$ is the limiting value of the bulk configurational stress $\mathbf{C}$, when the external boundary $\partial\mathrm{B}$ is approached from outside the body (i.e. $\Omega\backslash\mathrm{B}$). For regions of the boundary $\partial\mathrm{B}$, where tractions are prescribed, they are induced by air or an incompressible fluid. Where displacements are specified, they are assumed to be transferred by a rigid body. Consequently $\mathbf{C}^{\mathrm{out}} \equiv \mathbf{0}$, and Eq. (14) follows

$$\mathbf{f}_\mathrm{S} = \mathbf{C}^{\mathrm{in}}\hat{\mathbf{m}} = \phi^{\mathrm{in}}\hat{\mathbf{m}} - \left(\mathbf{F}^{\mathrm{in}}\right)^{\mathrm{T}}\mathbf{S}\hat{\mathbf{m}} \\ = \phi^{\mathrm{in}}\hat{\mathbf{m}} - \left(\mathbf{F}^{\mathrm{in}}\right)^{\mathrm{T}}\mathbf{t}. \tag{15}$$

The (stress- or) traction vector $\mathbf{t} = \mathbf{S}\hat{\mathbf{m}}$ is the applied traction on parts of $\partial\mathrm{B}$ where traction is prescribed, or has to be determined from the solution of the equilibrium problem for regions where displacement is prescribed.

3 The surface configurational forces and J-integrals

We now determine the relation between the surface configurational forces on the external boundary $\partial\mathrm{B}$ and the J-integrals used in fracture mechanics. Consider the gray region D in Fig. 3, which contains the entire external boundary of the body, including the crack faces, but not the crack-tip region. The balance of configurational forces for the gray region D requires

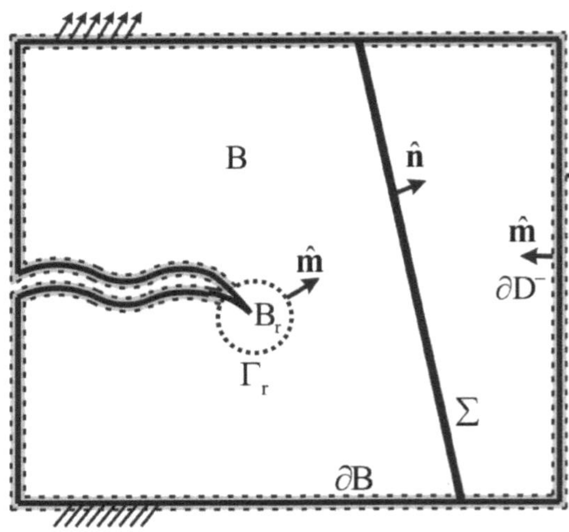

Fig. 3 The gray region D contains the entire external boundary of the body, including the crack faces, but not the crack-tip region. This region is used for relating the J-integral vector in the far-field and the surface configurational forces, Eq. (22)

$$\int_{D\cap(\Omega\backslash B)} \mathbf{f}\, dA + \int_{D\cap B} \mathbf{f}\, dA + \int_{\partial B} \mathbf{f}_S dl + \int_{\partial D} \mathbf{C}\hat{\mathbf{m}}\, dl = \mathbf{0}, \quad (16)$$

where the part of the gray region outside the fracture specimen is $D \cap (\Omega\backslash B)$, the part inside is $D \cap B$, and the traction due to the bulk configurational force acts on the boundary ∂D of the gray region. Next suppose that we collapse the gray region D in Fig. 3 to the external boundary ∂B. Then the first two integrals in Eq. (16) vanish, and the last integral can be written as

$$\int_{\partial D} \mathbf{C}\hat{\mathbf{m}}\, dl = \int_{\partial D^+} \mathbf{C}^{\text{out}}\hat{\mathbf{m}}\, dl + \int_{\partial D^-} \mathbf{C}^{\text{in}}\hat{\mathbf{m}}\, dl = \int_{\partial B} \mathbf{C}^{\text{out}}\hat{\mathbf{m}}\, dl - \int_{\partial B} \mathbf{C}^{\text{in}}\hat{\mathbf{m}}\, dl, \quad (17)$$

where the boundary $\partial D = \partial D^+ \cup \partial D^-$; the external normal to ∂D^- is opposite to the external normal on ∂B, hence the $-$ sign appears in front of the contour integral of $\mathbf{C}^{\text{in}}\hat{\mathbf{m}}$ on ∂B. Now Eq. (17) reduces to

$$\mathbf{J}_{\text{far}} - \mathbf{J}_{\text{out}} = \int_{\partial B} \mathbf{f}_S dl, \quad (18)$$

where we have used the following formal definitions of the J-integral vectors

$$\mathbf{J}_{\text{far}} = \int_{\partial B} \mathbf{C}^{\text{in}}\hat{\mathbf{m}}\, dl, \quad (19)$$

$$\mathbf{J}_{\text{out}} = \int_{\partial B} \mathbf{C}^{\text{out}}\hat{\mathbf{m}}\, dl. \quad (20)$$

If the region outside the body contains either vacuum, air, rigid material, or incompressible fluid, then Eq. (13) says that the bulk configurational force vanishes $\left(\mathbf{C}^{\text{out}} = \mathbf{0}\right)$; the only situation where this would not be true is the case, when the external region contains an inhomogeneous compressible fluid. Consequently for almost all cases encountered in structural applications, the J-integral vector in the external region vanishes identically,

$$\mathbf{J}_{\text{out}} = \mathbf{0}, \quad (21)$$

and we obtain the following relation between the J-integral vector in the far-field and the surface configurational forces

$$\mathbf{J}_{\text{far}} = \int_{\partial B} \mathbf{f}_S dl. \quad (22)$$

This means that the far-field J-integral vector can be obtained by integrating the surface configurational force over the entire external boundary of the body, which includes the crack faces, $\partial B = \Gamma_{\text{far}} + \Gamma^{\text{u}} + \Gamma^{\text{l}}$, but excludes the crack tip. We can readily recover the usual expression for the J-integral vector using Eq. (15) as

$$\mathbf{J}_{\text{far}} = \int_{\partial B} \mathbf{C}^{\text{in}}\hat{\mathbf{m}}\, dl = \int_{\partial B} \left[\phi^{\text{in}}\mathbf{I} - \left(\mathbf{F}^{\text{in}}\right)^{\text{T}}\mathbf{S}\right]\hat{\mathbf{m}}\, dl = \int_{\partial B} \left[\phi^{\text{in}}\hat{\mathbf{m}} - \left(\mathbf{F}^{\text{in}}\right)^{\text{T}}\mathbf{t}\right] dl. \quad (23)$$

The open literature and established software, such as ABAQUS (http://www.simulia.com/products/abaqus_fea.html), offer standard methods to evaluate the far-field J-integral as a scalar quantity, i.e. as projection of the J-integral vector $\mathbf{J}_{\text{far}}$ into the direction of crack propagation, $J_{\text{far}} = \mathbf{J}_{\text{far}}\cdot\mathbf{e}$, where $\mathbf{e}$ denotes the unit vector in crack growth direction, see below. Note that when computing the far-field J-integral with ABAQUS, not the outer contour ∂B is taken for the evaluation, but a contour Γ_{far} along the inward nodes of the elements situated along the outer boundary of the body. Moreover, Eq. (23) is not directly used; the J-integral values are calculated using the virtual crack extension method where the contour integral is transformed into an area integral (Parks 1977). Equation (22) provides a different, independent way of evaluating the far-field J-integral, if the surface configurational forces are provided by a postprocessor, e.g. Mueller et al. (2002, 2004), Shan (2008).

Lastly, we consider the region $B\backslash B_r$, assume a contour Γ_r at distance r from the crack tip, and form the limit $r \to 0$; this means finally the total region B excluding the crack tip. Now the balance of configurational forces reads

$$\int_{B\backslash B_r} \mathbf{f}\, dA + \int_{\Sigma} \mathbf{f}_\Sigma dl + \int_{\partial B} \mathbf{C}\hat{\mathbf{m}}\, dl + \int_{\Gamma_r} \mathbf{C}\left(-\hat{\mathbf{m}}\right) dl = \mathbf{0}. \quad (24)$$

The near-tip J-integral vector is written as

$$\mathbf{J}_{\text{tip}} = \lim_{r\to 0} \int_{\Gamma_r} \mathbf{C}\hat{\mathbf{m}}\, dl = -\mathbf{f}_{\text{T}}, \quad (25)$$

where $\mathbf{f}_{\text{T}}$ denotes the configurational force at the crack tip, Eq. (12). Then Eqs. (23)–(25) provide the relation between the far-field and near-tip J-integral vectors for $r \to 0$ in the form

$$\mathbf{J}_{\text{tip}} - \mathbf{J}_{\text{far}} = \int_{B} \mathbf{f} dA + \int_{\Sigma} \mathbf{f}_{\Sigma} dl. \tag{26}$$

Since the crack faces are included in the far-field J-integral vector, the surface configurational forces on the crack faces do not appear explicitly in Eq. (26). Consequently, we get the important result that the surface configurational forces on crack faces are irrelevant to the difference between the near-tip and far-field J-integral vectors. However, in the formulation of Eq. (22) and, consequently, Eq. (26), the surface ∂B includes the crack surfaces but excludes the crack tip.

Finally, it is recalled from Simha et al. (2003, 2005, 2008) that the product $\mathbf{J}_{\text{tip}} \cdot \mathbf{v}_{\text{tip}}$ is the dissipation due to the crack tip moving with a velocity $\mathbf{v}_{\text{tip}}$. The crack driving force is the projection of the vector $\mathbf{J}_{\text{tip}}$ into the direction of $\mathbf{v}_{\text{tip}}$, i.e. the scalar quantity $J_{\text{tip}} = \mathbf{J}_{\text{tip}} \cdot \mathbf{e}$, where the unit vector $\mathbf{e} = \mathbf{v}_{\text{tip}}/|\mathbf{v}_{\text{tip}}|$ lies along the direction of crack growth.

4 Surface configurational forces on crack faces

4.1 General relations for the far-field J-integral vector

Intuitively, we expect the far-field J-integral vector $\mathbf{J}_{\text{far}}$ to be related to the applied loads in the far-field. Since loads can also be applied to the crack faces as in hydraulic fracturing or by wedge-loading at the crack mouth, the far-field contour would have to be the entire external boundary of a body. In particular, the far-field contour should include the crack faces as defined in Eqs. (20), (22) and (26). In contrast, typically the crack faces are not included in the far-field contours Γ_{far} for the evaluation of the far-field J-integral vector $\mathbf{J}_{\text{far}}$ or the scalar far-field J-integral J_{far}. To clarify this we rewrite Eq. (22) as

$$\mathbf{J}_{\text{far}} = \int_{\partial B} \mathbf{f}_S dl = \int_{\Gamma_{\text{far}}} \mathbf{f}_S dl + \int_{\Gamma^{u}} \mathbf{f}_S dl + \int_{\Gamma^{l}} \mathbf{f}_S dl. \tag{27}$$

Only if the integrals along the upper and lower crack faces Γ^{u} and Γ^{l} vanished, then it would be sufficient to use a contour like Γ_{far} that excludes the crack faces. Consequently, we now examine the value of the surface configurational forces on the crack faces for several scenarios below.

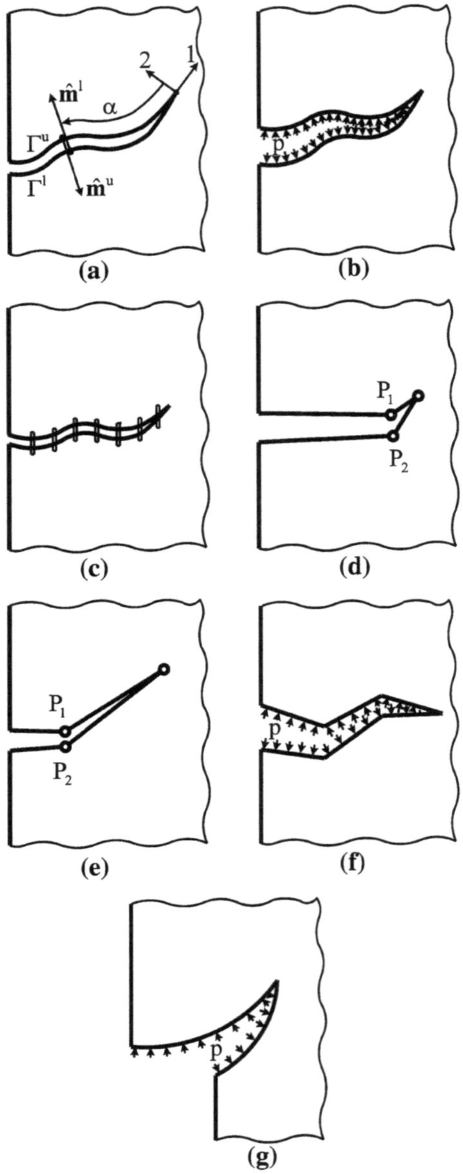

Fig. 4 Various crack configurations discussed in Sect. 4 for the examination of the surface configurational forces on the crack faces

4.2 Smooth and traction-free crack faces

Figure 4 shows a compilation of several representative crack configurations. Typically, the external unit normal vector to the upper crack face $\hat{\mathbf{m}}^{u}$ is opposite to the unit normal vector to the lower face $\hat{\mathbf{m}}^{l}$, which means $\hat{\mathbf{m}}^{l} = -\hat{\mathbf{m}}^{u}$ for a pair of opposing points that have the same arc-length from the crack tip, see Fig 4a. On a free crack surface there is also no traction (either applied or reactions) present on the two crack faces $(\mathbf{t}^{\mathrm{u}} = \mathbf{t}^{\mathrm{l}} = \mathbf{0})$. Under these circumstances, the surface configurational forces, Eq. (15), simplify to

$$\mathbf{f}_{\mathrm{S}} = \phi^{\mathrm{u}}\hat{\mathbf{m}}^{\mathrm{u}} \quad \text{on } \Gamma^{\mathrm{u}} \quad \text{and} \quad \mathbf{f}_{\mathrm{S}} = \phi^{\mathrm{l}}\hat{\mathbf{m}}^{\mathrm{l}} \quad \text{on } \Gamma^{\mathrm{l}}. \tag{28}$$

In the context of linear elastic materials and for straight cracks (straight at least near the crack tip with the local coordinates (1, 2) in Fig. 4a), all components of the singular crack-tip stress vanish, $\sigma_{11} = \sigma_{12} = \sigma_{22} = 0$, on both the upper and lower crack faces for Mode I cracks, see e.g. Gross and Seelig (2006), Anderson (2004). Consequently, also the strain energy densities on the upper and lower crack faces vanish, i.e. $\phi^{\mathrm{u}} = \phi^{\mathrm{l}} = 0$ in the near-tip region for Mode I cracks. For Mode II cracks σ_{11} is equal and opposite for opposing points on the upper and lower crack faces (the stress components $\sigma_{12} = \sigma_{22} = 0$), but the strain energy densities are equal, i.e. $\phi^{\mathrm{u}} = \phi^{\mathrm{l}} \neq 0$. Similarly, for Mode III cracks, the out-of-plane shear stresses are equal and opposite on opposing points on the upper and lower crack faces, but the strain energy densities will be equal.

As a conclusion, for cracks with traction-free crack faces in elastic materials we obtain the following results for the surface configurational force $\mathbf{f}_{\mathrm{S}}$:

- For Mode I cracks, $\mathbf{f}_{\mathrm{S}} = \mathbf{0}$ identically everywhere on the upper and lower crack faces both in the near-tip region and in the far-field, so the contribution from each of the crack faces vanishes, i.e. $\int_{\Gamma^{\mathrm{u}}} \mathbf{f}_{\mathrm{S}} dl = \int_{\Gamma^{\mathrm{l}}} \mathbf{f}_{\mathrm{S}} dl = \mathbf{0}$. If the non-singular stresses have some non-zero components (e.g. if non-zero T-stresses occur in bend-type fracture mechanics specimens with small crack lengths or center cracked tension specimens, (Williams 1957; Anderson 2004), then $\mathbf{f}_{\mathrm{S}}$ is equal and opposite in sign at opposing points on the upper and lower crack faces, yielding

$$\int_{\Gamma^{\mathrm{u}}} \mathbf{f}_{\mathrm{S}} dl = -\int_{\Gamma^{\mathrm{l}}} \mathbf{f}_{\mathrm{S}} dl \neq \mathbf{0}. \tag{29}$$

- For Mode II and Mode III cracks, $\mathbf{f}_{\mathrm{S}}$ is equal and opposite in sign at opposing points on the upper and lower crack faces in the near-tip region and decays to a zero vector in the far-field. In any case, the contribution from each of the faces is equal and opposite in sign, so again Eq. (29) will be valid.

Consequently, the net contribution of the surface configurational forces from both crack faces vanishes in linear elastic materials for all three modes, since

$$\int_{\Gamma^{\mathrm{u}}+\Gamma^{\mathrm{l}}} \mathbf{f}_{\mathrm{S}} dl = \int_{\Gamma^{\mathrm{u}}} \mathbf{f}_{\mathrm{S}} dl + \int_{\Gamma^{\mathrm{l}}} \mathbf{f}_{\mathrm{S}} dl = \mathbf{0}. \tag{30}$$

In elastic-plastic materials, both the slip line solution and the Hutchinson-Rice-Rosengren (HRR) stress field near the crack tip have a non-zero stress σ_{11} on the crack faces for Mode I cracks, see e.g. Hutchinson (1968), Rice and Rosengren (1968) and Gross and Seelig (2006), Sect. 5.3 there. However, the strain energy density is the same at opposing points on the upper and lower crack faces. Consequently, Eqs. (29) and (30) will be valid also for Mode I cracks in elastic-plastic materials. Similar investigations can be performed for Mode II in elastic-plastic materials, but need a further investigation, being not the topic of this paper.

4.3 Smooth but non-zero traction crack faces

The traction vectors could either be applied traction vectors like in hydraulic fracturing (Fig. 4b), or reaction forces due to local contact of the crack faces, such as crack closure during cyclic loading, or due to crack-bridging fibers in composites (Fig. 4c). For many such cases, the traction will be equal but opposite in sign at opposing points on the upper and lower crack faces, i.e. $\mathbf{t}^{\mathrm{u}} = -\mathbf{t}^{\mathrm{l}}$. For symmetric or antimetric loading and a symmetric geometry with respect to the crack (i.e. crack lies in X-direction, normals to crack faces in Y-direction, Z-direction in out-of-plane direction; corresponding unit vectors $\mathbf{e}_X, \mathbf{e}_Y, \mathbf{e}_Z$; corresponding displacement vector $\mathbf{u} = u\mathbf{e}_X + v\mathbf{e}_Y + w\mathbf{e}_Z$), the resultant configurational force in two opposing points of the crack faces acts for Mode I and II in the crack-direction and for Mode III in the out-of-plane direction as

$$\text{Mode I } \mathbf{t}^{\mathrm{u}} = t\mathbf{e}_Y\colon\ \mathbf{f}_{\mathrm{S}}^{\mathrm{u}} + \mathbf{f}_{\mathrm{S}}^{\mathrm{l}} = -2t\frac{\partial v^{\mathrm{u}}}{\partial X}\mathbf{e}_X, \tag{31.1}$$

$$\text{Mode II } \mathbf{t}^{\mathrm{u}} = t\mathbf{e}_X\colon\ \mathbf{f}_{\mathrm{S}}^{\mathrm{u}} + \mathbf{f}_{\mathrm{S}}^{\mathrm{l}} = -2t\frac{\partial u^{\mathrm{u}}}{\partial X}\mathbf{e}_X, \tag{31.2}$$

$$\text{Mode III } \mathbf{t}^{\mathrm{u}} = t\mathbf{e}_Z\colon\ \mathbf{f}_{\mathrm{S}}^{\mathrm{u}} + \mathbf{f}_{\mathrm{S}}^{\mathrm{l}} = -2t\frac{\partial w^{\mathrm{u}}}{\partial Z}\mathbf{e}_Z. \tag{31.3}$$

The parameter t in Eq. (31) denotes the magnitude of the traction per length unit of the crack faces and unit thickness. These relations are also valid for cracks in elastic-plastic materials.

4.4 Non-smooth and non-zero traction crack faces

There are a few situations where Eq. (30) may not hold, as in the following:

- Traction-free crack faces but with kinks on the crack faces close to the crack tip (Fig. 4d). There can be an additional singularity at some of these kinks, see e.g. in P_2, Fig. 4d for a material angle $> \pi$, and Simha and Bhattacharya (1998), which would then appear as a non-zero surface configurational force. In general, $\mathbf{f}_S$ will have different values at kinks on the upper and lower faces and, hence, Eq. (30) may not be valid. The singular stress field at such a kink will likely influence the near-tip stress fields as well. However, if the kinks are in the far-field, the surface configurational force will likely be negligibly small (Fig. 4e).
- Applied tractions on crack faces plus kinks on crack faces (Fig. 4f). The stress field at some of these kinks could be singular, resulting in unequal, non-zero surface configurational forces on the upper and lower crack faces. Notice that any singularity at the crack mouth (like in the case of loading by a wedge at the crack mouth) would also fall under this case.
- Asymmetrical cracks plus applied tractions on crack faces (Fig. 4g): By asymmetry, we mainly mean that the consequence will be $\mathbf{F}^u \neq \mathbf{F}^l$. This could be the case, if there is a mixed-mode loading on the crack face; for a specific treatment, see e.g. Sect. 4.3 of the current paper and Kienzler et al. (2009).

However, for all these cases Eq. (30) may hold approximately, in the sense that the integrals in Eq. (30) may be vanishingly small in comparison to the magnitude of $\mathbf{J}_{far}$ for instance; this will be verified in future computational studies.

Some final remarks shall be made at the end of this section. For every case when Eq. (30) is valid, Eq. (27) reduces to

$$\mathbf{J}_{far} = \int_{\partial B} \mathbf{f}_S dl = \int_{\Gamma_{far}} \mathbf{f}_S dl. \tag{32}$$

Consequently, the current consensus practice of calculating the far-field J-integral on the contour Γ_{far} is valid for the vast majority of applications.

We consider it as important to note that the current derivation does not include any constitutive relation.

5 Numerical studies

Computational studies are conducted in order to evaluate (i) the relation between the far-field J-integral vector and the surface configurational forces, Eq. (22), and (ii) the influence of tractions on crack faces on the path dependence of the J-integral vector.

The modeling and computation of stresses, strains and energy densities are performed with ABAQUS (http://www.simulia.com/products/abaqus_fea.html). The configurational forces are calculated by a special post-processing procedure, see Simha et al. (2005, 2008), Kolednik et al. (2010), Shan (2008), Mueller et al. (2002, 2004).

5.1 Conventional fracture mechanics specimen, elastic-plastic material

As first example we investigate a homogeneous Compact Tension specimen with crack length $a = 27$ mm, width $W = 50$ mm, height $2H = 60$ mm, compare Fig. 5. The material is a homogeneous elastic-plastic material with Young's modulus $E = 200$ GPa, Poisson's ratio $\nu = 0.3$, yield strength $\sigma_y = 270$ MPa, ultimate tensile strength $\sigma_u = 426$ MPa, and strain hardening exponent $n = 0.2$, similar to a mild steel with German designation St37. Plane strain conditions are assumed for the computation. The finite element mesh consists of two-dimensional 4-node isoparametric elements. The elastic-plastic material behavior is modeled using the incremental plasticity model provided by ABAQUS. The specimen is loaded in Mode I by prescribing the displacement v_{LL} at the load application points (marked as LP in Fig. 5a) at the left boundary; no other traction forces exist. The material near the load application points is modeled as linear-elastic in order to avoid large artificial strains. Note that the specimen behaves as a homogeneous body as long as the plastically deformed zone does not touch this elastic region. The configurational forces are evaluated as outlined in Simha et al. (2008), Schöngrundner et al. (2010).

The distribution of the configurational forces for $v_{LL} = 0.25$ mm is depicted in Fig. 5. Note that somewhat modified configurational forces $\tilde{\mathbf{f}}_s$ are depicted. The reason is that considerable numerical errors may appear when calculating the configurational forces $\mathbf{f} = -\nabla|_{expl} \cdot \mathbf{C} = -\nabla|_{expl} \cdot (\phi\mathbf{I} - \mathbf{F}^T\mathbf{S})$, Eq. (9), from the deformation gradient $\mathbf{F} = \mathbf{I} + \nabla\mathbf{u}$ for small

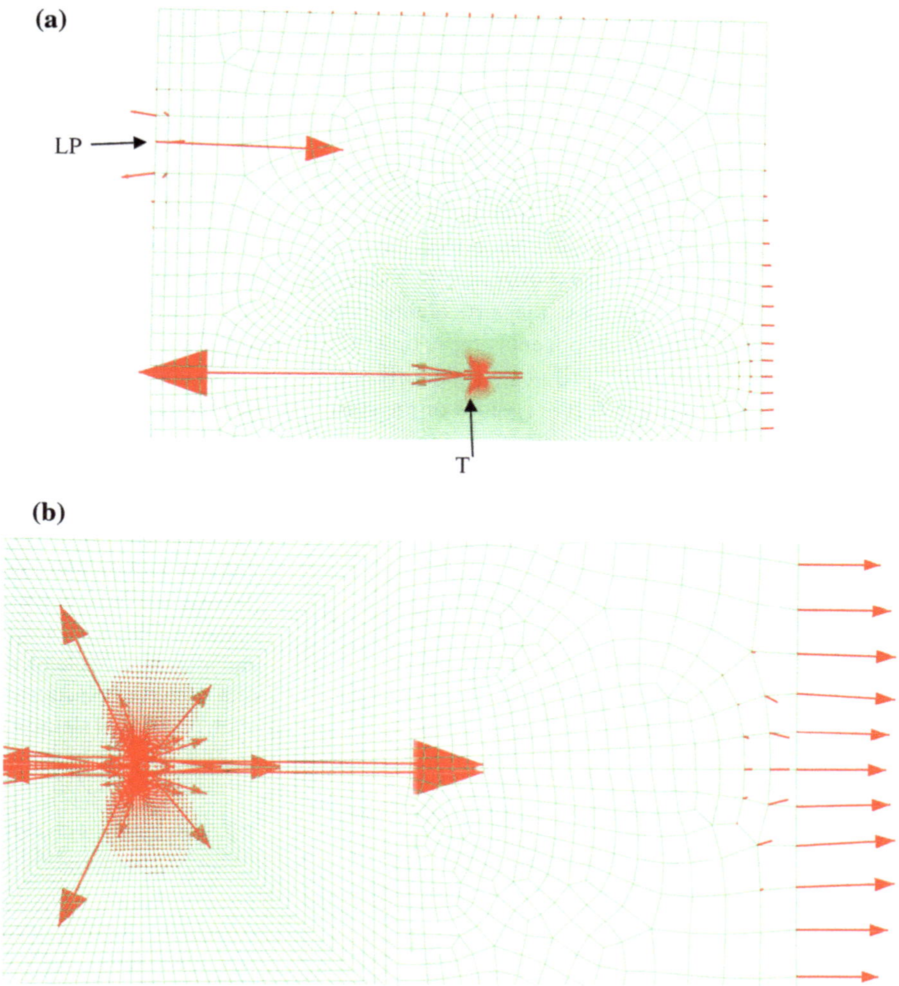

Fig. 5 Distribution of surface configurational forces $\tilde{\mathbf{f}}_s$ at the boundaries (*red arrows*) and configurational body forces in the interior (*red lines*) of a Compact Tension specimen. A large configurational force appears also at the crack tip T. **a** Overview of whole specimen; note the large horizontal configurational force at the load application point LP. **b** Detail near the crack tip

deformations. To avoid this, modified configurational forces are evaluated in the form $\tilde{\mathbf{f}} = -\nabla|_{\text{expl}} \cdot \tilde{\mathbf{C}} = -\nabla|_{\text{expl}} \cdot (\phi \mathbf{I} - \nabla \mathbf{u}^T \mathbf{S})$ yielding $\nabla|_{\text{expl}} \cdot \tilde{\mathbf{C}} = \nabla|_{\text{expl}} \cdot \mathbf{C}$ everywhere in the body. The only difference appears regarding the surface configurational forces at the load application points where large traction vectors $\mathbf{t}$ appear. Then $\tilde{\mathbf{f}}_s = \mathbf{f}_s + \mathbf{t}$, and so the large vertical (and opposite) traction vectors in the load application points are omitted when plotting $\tilde{\mathbf{f}}_s$ instead of $\mathbf{f}_s$ in Fig. 5.

Bulk configurational forces appear due to the gradient of plastic strain within the plastic zone near the crack tip and near the back face of the specimen (Simha et al. 2008). Additional bulk configurational forces are visible near the load application points, but these are caused by numerical inaccuracies due to high local strains. Note that within a homogeneous elastic material no configurational body forces are expected. Surface configurational forces are present on the upper, right, lower and left parts of the external boundary, as well as on the crack faces. The surface configurational forces on the upper and lower boundaries and on the crack faces are small; they have no components in the horizontal 1-direction. In particular, notice that the values on the upper and lower crack faces are equal and opposite on a pair of opposing points on the two crack faces with the same distance from the crack tip.

Large surface configurational forces appear on the left boundary near the load application points and on the right boundary; in both cases, the components in the horizontal (1-) direction are considerably larger than those in the vertical (2-) direction. Together they deliver the main contribution to the total far-field J-integral vector, which has a non-zero component in the horizontal direction. The vertical components of the surface configurational forces sum up to zero; hence the far-field J-integral vector has a zero vertical component.

These findings can be also deduced from Fig. 6, which presents the various contributions to the far-field J-integral vector $\mathbf{J}_{\partial B}$. Figure 6a collects the contributions with respect to the horizontal (1-) direction, Fig. 6b the contributions to the vertical direction. Shown are the J-integral components originating from the boundaries on the left ($\partial B - \ell$), right ($\partial B - r$), upper ($\partial B - u$) and lower ($\partial B - l$) sides of the specimen, as well as the contributions from the upper and lower crack faces, (CF − u) and (CF − l). Shown are, in addition, the near-tip J-integral vector $\mathbf{J}_{tip}$, which equals the negative configurational force at the crack tip $-\mathbf{f}_T$ (Eq. 25), the far-field J-integral vector $\mathbf{J}_{\partial B}$ evaluated from the whole boundary ∂B of the body (Eq. 22), as well as the scalar J-integral values $J_{\Gamma 3}$ and J_{far}, which are calculated along the path with a distance of 3 elements from the crack tip for Γ_3, i.e. a near-tip path, and the far-field path Γ_{far}, i.e. along the inner nodes of the outer boundary elements.

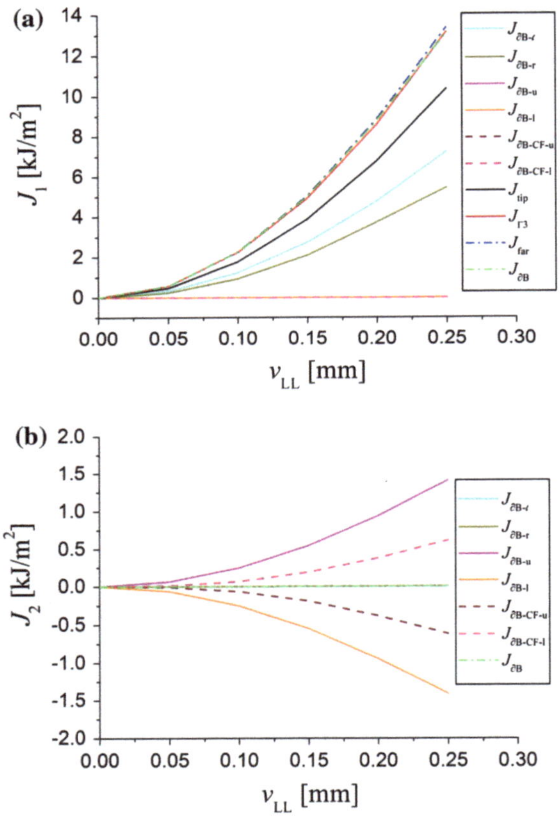

Fig. 6 Components of the J-integral vectors in **a** the horizontal (1-) direction and **b** the vertical (2-) direction, plotted against the load line displacement v_{LL}. The J-integral vectors are evaluated along the whole boundary ∂B of the body and along parts of the boundary on the *left* ($\partial B - \ell$), *right* ($\partial B - r$), *upper* ($\partial B - u$) and *lower* ($\partial B - l$) side, as well as on the *upper* and *lower* crack faces, (CF − u) and (CF − l). The near-tip J-integral vector $\mathbf{J}_{tip}$ equals the negative configurational force at the crack tip $-\mathbf{f}_T$, Eq. (25). $J_{\Gamma 3}$ and J_{far} are the values of the conventional, scalar J-integrals along the paths Γ_3 and Γ_{far}

An important observation in Fig 6a is that the scalar far-field J-integral, i.e. the projection of $\mathbf{J}_{\partial B}$ into the direction of the crack, calculated from the surface configurational forces according to Eq. (22), agrees closely with the standard contour integral J_{far} from ABAQUS. This provides evidence of the high accuracy of the numerical computations. The reader is also referred to the fact that the crack driving force J_{tip} differs from the scalar, far-field J-integral J_{far} due to plastic straining and incremental plasticity, which causes the occurrence of a non-zero plasticity influence term C_p, see Simha et al. (2008) for details.

5.2 Specimens with different loading cases, linear elastic material

As second example, we consider again a Compact Tension specimen ($a = 29$ mm, $W = 50$ mm, $2H = 60$ mm). The material behaves as linear elastic with $E = 210$ MPa and $\nu = 0.3$. Again plane strain conditions are assumed for the computations. Three different loading Cases are studied, namely

(i) Mode II loading by constant pressure $p = 100$ MPa between the crack faces;
(ii) Mode I loading by variable pressure between the crack faces, $p = p_{max}(1 - X/a)$ with $p_{max} = 200$ MPa; X is measured from the crack mouth towards the crack tip;
(iii) Mode II loading by constant shear stress $\tau = 100$ MPa on the upper and -100 MPa on the lower crack face.

Since neither a material inhomogeneity nor an elastic-plastic behavior exists, no configurational body forces appear and the J-integral vector is path independent, $\mathbf{J}_{far} = \mathbf{J}_{tip}$.

In the following, analytical expressions of the far-field J-integral vectors $\mathbf{J}_{\text{far}}$ shall be derived according to Eq. (23) for the three loading cases, compare also Eq. (27). No contribution to $\mathbf{J}_{\text{far}}$ arises from the upper ("u") and the lower ("l") boundaries, since no traction vector is acting there and since, due to symmetry, the elastic strain energy density ϕ has identical values, but opposite normal vectors $\hat{\mathbf{m}}$ at corresponding points on "u" and "l". No traction vectors are present also on the left ("ℓ") and right ("r") boundaries of the specimen, but the values of the elastic strain energy density ϕ will be significantly different at corresponding points on "ℓ" and "r", with opposite signs of the unit normal vectors $\hat{\mathbf{m}}$. The contributions of the upper and lower crack faces to $\mathbf{J}_{\text{far}}$ follow in straightforward way from Eqs. (31.1) and (31.2). With these considerations, the far-field J-integral vectors $\mathbf{J}_{\text{far}}$ follow as

$$\mathbf{J}_{\text{far}} = \left[\left(\int_{\text{r}} \phi ds - \int_{\ell} \phi ds \right) - 2 \int_{\Gamma^{\text{u}}} p\,(X) \frac{dv^{\text{u}}}{dX} dX \right] \mathbf{e}_X \tag{33}$$

for the Cases (i) and (ii) for constant and variable pressure and Mode I loading, resp., or as

$$\mathbf{J}_{\text{far}} = \left[\left(\int_{\text{r}} \phi ds - \int_{\ell} \phi ds \right) - 2 \int_{\Gamma^{\text{u}}} \tau \frac{du^{\text{u}}}{dX} dX \right] \mathbf{e}_X \tag{34}$$

for the Case (iii) for constant shear and Mode II loading.

Evaluation of the integrals Eqs. (33) and (34) yields after some analysis the following results for the X- and Y-components of the far-field J-integral vectors:

(i) constant pressure, Mode I

$$J_{\text{far},X} = \int_{-H}^{+H} (\phi_{\text{r}} - \phi_{\ell})\, dY + 2 p v_{\text{mouth}}, \qquad J_{\text{far},Y} = 0; \tag{35}$$

(ii) variable pressure, Mode I

$$J_{\text{far},X} = \int_{-H}^{H} (\phi_{\text{r}} - \phi_{\ell})\, dY + 2 p_{\max} (v_{\text{mouth}} - \bar{v}), \qquad J_{\text{far},Y} = 0; \tag{36}$$

(iii) constant shear, Mode II

$$J_{\text{far},X} = \int_{-H}^{H} (\phi_{\text{r}} - \phi_{\ell})\, dY + 2 \tau\, u_{\text{mouth}}, \qquad J_{\text{far},Y} = 0. \tag{37}$$

In the equations above, u_{mouth} and v_{mouth} are the displacements of the upper crack face at the mouth of the crack, i.e., at $X = 0$, in horizontal (X- or 1-) and vertical (Y- or 2-) direction. The average value of the crack mouth displacement at the upper crack face in vertical direction is denoted in Eq. (36) as $\bar{v}$.

It is interesting to consider the contributions of the crack faces for a concrete numerical example for each of the three loading cases presented above. To do so, ABAQUS was used to calculate the deformation and stress state and the elastic strain energy ϕ. The integrals in Eqs. (35), (36) and (37) were then evaluated numerically by a postprocessor. The FE-computations were performed for both small and large strain setting, with practically no difference in the results. Therefore, only the results for small strain setting are used in the following. The results are collected in Table 1 under "Semi-analytical Solutions"; only the J-integral components in X-direction are listed. The term $J_{\partial\text{B}-\text{r}}$ denotes the contribution of the right vertical boundary to the far-field J-integral. $(J_{\partial\text{B}-\text{l}\ell} + J_{\partial\text{B}-\text{u}\ell})$ is the total contribution of the left boundary; hereby the contributions of the lower and upper specimen half are given separately. The term $(J_{\partial\text{B}-\text{CF}-\text{u}} + J_{\partial\text{B}-\text{CF}-\text{l}})$ denotes the contribution from the upper and lower crack faces.

As second way, the configurational force postprocessor to ABAQUS, Shan (2008), was engaged to calculate the distribution of the configurational forces along the different parts of the boundary of the specimen and calculate J_{far} from Eq. (27). The results are collected in Table 1 under "Numerical solutions". In addition, the standard numerical solutions of the J-integral J^{VCE} provided by the ABAQUS virtual crack extension procedure are listed in Table 1. Note that for elastic homogeneous materials the near-tip J-integral equals the far-field value, $J_{\text{tip}} = J_{\text{far}}$.

Figure 7 shows an overview and some details of the configurational force distribution for Case 2, i.e. the loading by variable pressure between the crack faces. The image in the center gives an overview about the whole specimen. Above are shown the configurational forces near the crack tip, below the configurational forces along the crack faces, and on the right the configurational forces at the right boundary of the specimen. Figure 8 presents the contributions to the far-field J-integral VCE vector from the various parts of the boundary. Figure 8a shows the components of the J-integral vectors in horizontal (1-) direction, Fig. 8b those in the vertical (2-) direction. It might be inter-

Table 1 Contributions to the X-component $J_{far,X}$ of the far-field J-integral in kJ/m^2, resulting from various regions along the external boundary

Loading case	Semi-analytical solutions				Numerical solutions				ABAQUS J^{VCE}
	$J_{\partial B-r}$	$J_{\partial B-l\ell} + J_{\partial B-u\ell}$	$J_{\partial B-CF-u} + J_{\partial B-CF-l}$	$J_{far,X}$	$J_{\partial B-r}$	$J_{\partial B-l\ell} + J_{\partial B-u\ell}$	$J_{\partial B-CF-u} + J_{\partial B-CF-l}$	$J_{far,X}$	$J^{*}_{tip,X}$
Constant pressure	16.020	−0.334 − 0.334	200·0.200	55.352	17.096	−0.033 − 0.033	36.300	53.334	56.030
Variable pressure	26.840	−1.165 − 1.165	400 · (0.262 − 0.149)	69.710	28.646	−0.386 − 0.386	21.338+21.338	70.550	71.753
Constant shear	0.287	−0.593 − 0.593	200·0.043	7.701	0.292	−0.653 − 0.653	3.951+4.183	7.120	7.479

* $J_{tip,X} = J_{far,X}$.

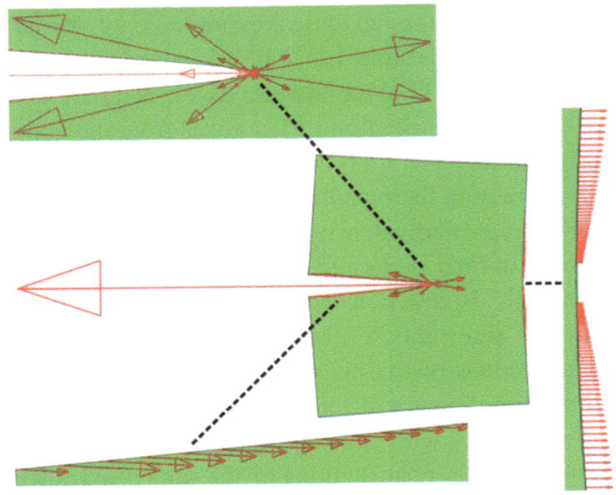

Fig. 7 Overview and some details of the distribution of the configurational forces for Case 2, i.e. loading by a linearly varying pressure p between the crack faces ($p_{max} = 200$ MPa at crack mouth, decreasing linearly to $p = 0$ at crack tip)

esting to compare these curves with the results listed in the second line of Table 1 under "Semi-analytical Solutions".

The results listed in Table 1 allow a comparison between the analytical procedure (columns referring to "Semi-analytical Solutions") and the results from the configurational forces postprocessor (columns referring to "Numerical solutions") and can be checked with the results from the ABAQUS VCE-procedure (last column). One can conclude as follows:

- All three ways to calculate J_{far} lead to similar results within "usual technical accuracy".
- The "Semi-analytical solutions" can be considered as the most accurate values. They use both displacements and elastic strain energy terms "far away" from the crack tip and, therefore, show the lowest dependency on mesh size and other numerical effects.
- The crack faces deliver the largest contribution to J_{far}, the left vertical boundaries the smallest contribution.
- The role of nonlinear geometrical behavior (large strain setting) can be taken into account, but has nearly no influence on the results for the actual case.

Finally, the reader is referred to the paper by Fischer et al. (2007), where (semi-) analytical J-integral solutions were studied, too, for several crack configurations, but without loading along the crack faces.

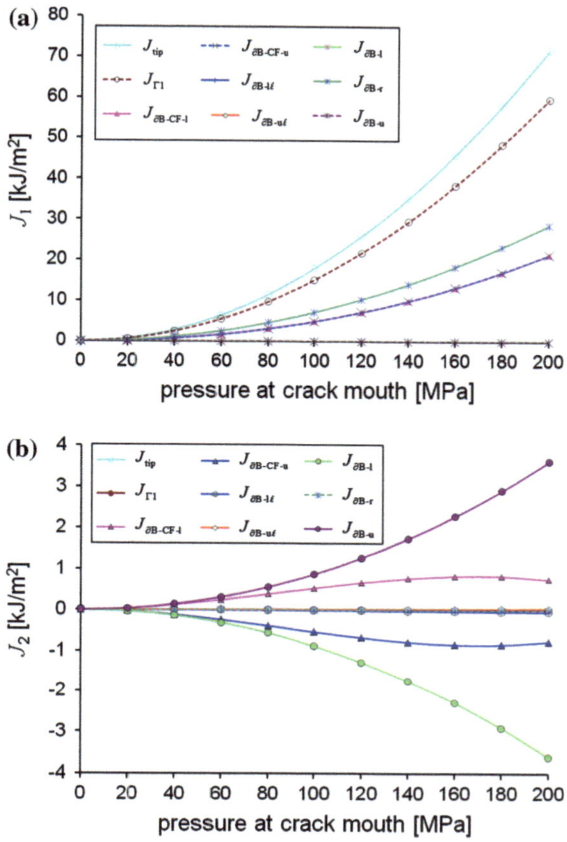

Fig. 8 Components of the J-integral vectors in **a** the horizontal (1-) direction and **b** the vertical (2-) direction. The J-integral vectors are evaluated at the crack tip $\left(\mathbf{J}_{\text{tip}} = \mathbf{J}_{\text{far}}\right)$ and along parts of the boundary on the *left* ($\partial \text{B} - \ell$), *right* ($\partial \text{B} - \text{r}$), *upper* ($\partial \text{B} - \text{u}$) and *lower* ($\partial \text{B} - \text{l}$) side as well as on the *upper* and *lower* crack flanks ($\partial \text{B} - \text{CF} - \text{u}$, $\partial \text{B} - \text{CF} - \text{l}$). The contribution of the boundary on the left side is split into two components resulting from the upper and lower specimen part, $J_{\partial \text{B}-\text{u}\ell}$ and $J_{\partial \text{B}-\text{l}\ell}$, resp. Note that J_{tip} is the negative configurational force at the crack tip. $J_{\Gamma 1}$ is the value of the conventional, scalar J-integral along path Γ_1

6 Summary

The results of this paper can be summarized as follows:

(1) Expressions for the surface configurational forces have been derived and their values have been determined on regions of the external boundaries with prescribed tractions or displacements.

(2) The far-field J-integral vector $\mathbf{J}_{\text{far}}$ can be obtained by integrating the surface configurational forces over the entire external boundary of the body, which includes the crack faces, but excludes the crack tip.

(3) Only some special cases exist where the surface configurational forces along the crack faces give a contribution to the far-field J-integral vector $\mathbf{J}_{\text{far}}$. In many cases the contributions of the upper and lower crack faces are zero or cancel.

(4) The surface configurational forces on the crack faces do not alter the relation between the near-tip J-integral vector $\mathbf{J}_{\text{tip}}$ and far-field J-integral vector $\mathbf{J}_{\text{far}}$, i.e. the J-integral is path independent for elastic homogeneous materials.

(5) In our numerical examples where the specimens are loaded along the crack faces, the ABAQUS VCE-procedure give the correct values of the scalar far-field J-integral J_{far}, however, there might exist some cases where this is not so.

Acknowledgments F.D.F. expresses his thanks to the valuable comments by Prof. H. Irschik, Johannes Kepler University Linz, Austria. Financial support by the Austrian Federal Government and the Styrian Provincial Government within the research activities of the K2 Competence Center "Integrated Research in Materials, Processing and Product Engineering", under the frame of the Austrian COMET Competence Center Programme, is gratefully acknowledged (Project A4.11-WP2).

References

Anderson TL (2004) Fracture mechanics: fundamentals and applications, 3rd edn. CRC Press, Boca Raton

Fischer FD, Predan J, Kolednik O, Simha NK (2007) Application of material forces to fracture of inhomogeneous materials: illustrative examples. Arch Appl Mech 77:95–112. Erratum (2008) 78:835

Gross D, Seelig T (2006) Fracture mechanics, with an introduction to micromechanics. Springer, Berlin

Gurtin ME, Podio-Guidugli P (1996) Configurational forces and the basic laws for crack propagation. J Mech Phys Solids 44:905–927

Gurtin ME (2000) Configurational forces as basic concepts of continuum physics. Springer, Berlin

Hutchinson JW (1968) Singular behavior at the end of a tensile crack tip in a hardening material. J Mech Phys Solids 16:13–31

Irschik H (2007) On rational treatments of the general laws of balance and jump, with emphasis on configurational formulations. Acta Mech 194:11–32

Kienzler R, Herrmann G (2000) Mechanics in material space. Springer, Berlin

Kienzler R, Rohde L, Schröder R, Kutz K (2009) Treating mixed-mode problems with path-independent integrals. In: Elboujdaini M et al. (eds) Proc 12th International Conference on Fracture (ICF12), paper T06.029, 10 p

Kolednik O, Predan J, Fischer FD (2010) Reprint of cracks in inhomogeneous materials: comprehensive assessment using the configurational forces concept. Eng Fract Mech 77:3611–3624

Maugin GA (1993) Material inhomogeneities in elasticity. Chapman & Hall, London

Maugin GA (1995) Material forces: concepts and applications. ASME Appl Mech Rev 48:213–245

Maugin GA (2010) Configurational forces: thermomechanics, physics, mathematics, and numerics. CRC Press, Boca Raton

Mueller R, Kolling S, Gross D (2002) On configurational forces in the context of the finite element method. Int J Numer Meth Eng 53:1557–1574

Mueller R, Gross D, Maugin GA (2004) Use of material forces in adaptive finite element methods. Comput Mech 33:421–434

Parks DM (1977) The virtual crack extension method for nonlinear material behavior. Comput Meth Appl Mech Eng 12:353–364

Rice JR (1968) A path independent integral and the approximate analysis of strain concentration by notches and cracks. ASME J Appl Mech 35:379–386

Rice JR, Rosengren GF (1968) Plane strain deformation near a crack tip in a power-law hardening material. J Mech Phys Solids 16:1–12

Schöngrundner R, Kolednik O, Fischer FD (2010) The configurational force concept in elastic-plastic fracture mechanics—instructive examples. Key Eng Mater 417-418:297–300

Shan GX (2008) Post-processing program for calculating the configurational forces with ABAQUS, internal report, K2 Competence Center "Integrated research in materials, processing and product engineering", Leoben, Austria

Simha NK, Bhattacharya K (1998) Kinetics of a phase boundary with edges and junctions. J Mech Phys Solids 46:2323–2359

Simha NK, Fischer FD, Kolednik O, Chen CR (2003) Inhomogeneity effects on the crack driving force in elastic and elastic-plastic materials. J Mech Phys Solids 51:209–240

Simha NK, Predan J, Kolednik O, Fischer FD, Shan GX (2005) Crack tip shielding or anti-shielding due to smooth and discontinuous material inhomogeneities. Int J Fract 135:73–93

Simha NK, Fischer FD, Shan GX, Chen CR, Kolednik O (2008) J-integral and crack driving force in elastic-plastic materials. J Mech Phys Solids 56:2876–2895

Steinmann P (2000) Application of material forces to hyperelastostatic fracture mechanics part I: continuum mechanical setting. Int J Solids Struct 37:7371–7391

Williams ML (1957) On the stress distribution at the base of a stationary crack. ASME J Appl Mech 24:109–114

Int J Fract (2012) 174:75–85
DOI 10.1007/s10704-011-9660-4

ORIGINAL PAPER

HotQC simulation of nanovoid growth under tension in copper

M. P. Ariza · I. Romero · M. Ponga · M. Ortiz

Received: 7 September 2011 / Accepted: 29 November 2011 / Published online: 20 December 2011

Abstract We apply the HotQC method of Kulkarni et al. (J Mech Phys Solids 56:1417–1449, 2008) to the study of quasistatic void growth in copper single crystals at finite temperature under triaxial expansion. The void is strained to 30% deformation at initial temperatures and nominal strain rates ranging from 150 to 600 K and from 2.5×10^5 to 2.5×10^{11} s^{-1}, respectively. The interatomic potential used in the calculations is Johnson's Embedded-Atom Method potential Johnson (Phys Rev B 37:3924–3931, 1988). The computed pressure versus volumetric strain is in close agreement with that obtained using molecular dynamics, which suggests that inertia effects are not dominant for the void size and conditions considered. Upon the attainment of a critical or cavitation strain of the order of 20%, dislocations are abruptly and profusely emitted from the void and the rate of growth of the void increases precipitously. Prior to cavitation, the crystal cools down due to the thermoelastic effect. Following cavitation dislocation emission causes rapid local heating in the vicinity of the void, which in turn sets up a temperature gradient and results in the conduction of heat away from the void. The cavitation pressure is found to be relatively temperature-insensitive at low temperatures and decreases markedly beyond a transition temperature of the order of 250 K.

Keywords Nanovoids · Cavitation · Dislocations · Quasicontinuum · Finite temperature · Copper · Ductile fracture

M. P. Ariza (✉) · M. Ponga
Escuela Técnica Superior de Ingeniería,
Universidad de Sevilla, Sevilla 41092, Spain
e-mail: mpariza@us.es

M. Ponga
e-mail: mponga@us.es

I. Romero
ETSI Industriales, Universidad Politécnica de Madrid,
Madrid 28006, Spain
e-mail: ignacio.romero@upc.es

M. Ortiz
Division of Engineering and Applied Science, California
Institute of Technology, Pasadena, CA 91125, USA
e-mail: ortiz@aero.caltech.edu

1 Introduction

Void nucleation, growth and coalescence are important micromechanical mechanisms underlying ductile crack growth in metals (Puttick 1959; Bauer and Wilsdorf 1973; Argon et al. 1975; Hancock and Mackenzie 1976; Johnson 1981; Meyers and Aimone 1983; Aoki et al. 1984; Garrison and Moody 1987). Multiscale models of ductile fracture often rely on porous plasticity in order to account for the macroscopic effect of void growth, and on micromechanical or localization analyses in order to describe crack growth of the formation of spall planes [cf., e.g., (Tvergaard 1990) and references therein]. Of necessity, continuum models of porous plasticity must either postulate an initial void size and density or rely on a nucleation model in order to compute the initial void size and density. In the present context, void nucleation specifically refers to the onset of plastic cavitation, i.e., to the development of voids of a size such that subsequent growth can take place by dislocation emission (Lubarda et al. 2004; Marian et al. 2004, 2005; Ahn et al. 2007; Meyers et al. 2009; Reina et al. 2011).

Molecular dynamics (MD) has been extensively used for the study of dislocation emission from nano-sized voids and their subsequent plastic growth (Belak 1998, 2002; Strachan et al. 2001; Rudd and Belak 2002; Seppälä et al. 2004a,b; Belak 2005; Dávila et al. 2005; Traiviratana et al. 2008; Bringa et al. 2010). While these studies reveal useful insights into the dislocation-mediated nanovoid cavitation and growth, they also suffer from the size and time limitations characteristic of MD. In particular, the need to track the thermal vibrations of the atoms places severe limitations on the time scales that are accessible to MD. This renders MD particularly well-suited—but also essentially limited—to the study of void growth at very high strain rates. In order to overcome these limitations, Marian et al. (2004, 2005) based their studies of dislocation emission and void growth on the Quasicontinuum (QC) method of Tadmor et al. (1996). Thus, by adapting the spatial resolution to the structure of the deformation field, the QC method facilitates consideration of realistic sample sizes representative of the conditions experienced by nanovoids in actual material samples. However, the calculations of Marian et al. (2004, 2005) were restricted to 0 K. While a number of finite-temperature extensions of QC have been proposed within the framework of equilibrium statistical mechanics and thermodynamics (Dupuy et al. 2005; Tang et al. 2006; Marian et al. 2010), these extensions are restricted to equilibrium conditions and, in particular, uniform temperature, and cannot be used to simulate the rapidly transient and highly non-equilibrium conditions attendant to dynamic void growth.

Kulkarni et al. (2008) have proposed a non-equilibrium extension of QC, referred to as HotQC, based on Jaynes' principle of maximum entropy (Jaynes 1957a,b) and on variational mean-field theory in order to construct effective thermodynamic potentials for systems of atoms away from equilibrium. In particular, the temperature field is not required to be uniform but, contrariwise, it is allowed to take different values from atom to atom. In addition, Kulkarni et al. (2008) resort to Yang et al. (2006) variational formulation of coupled transport problems in order to couple the non-equilibrium thermodynamic potentials to the heat equation. The chief advantage of Kulkarni et al. (2008) HotQC approach is that it entirely bypasses the need to track the thermal fluctuations of the atoms. In particular, the time steps required in calculations are set by the diffusional time scale, which is much greater than the time scale of the atomic vibrations.

In this paper, we apply the HotQC method to the study of quasistatic void growth in copper single crystals at finite temperature under triaxial expansion. We specifically consider a cubic computational cell of 78 nm in size encompassing 216^3 unit cells and containing a total of 5.1×10^6 atoms. The boundary of the cell is aligned with the cube directions. A spherical 2.0 nm void is introduced at the center of the cell and full atomistic resolution is initially provided within a $8a \times 8a \times 8a$ region. Away from this fully-atomistic region, the initial triangulation is gradually coarsened with the result that the initial computational mesh contains 900 representative atoms only. The mesh is subsequently adapted so as to ensure that all dislocations are embedded within a sizeable atomistic region at all times (Knap and Ortiz 2001). The computational cell is strained to 30% deformation by means of displacement control at its outer boundary. The initial temperatures and nominal strain rates under consideration range from 150 to 600 K and from 2.5×10^5 to 2.5×10^{11} s^{-1}, respectively. The interatomic potential used in the calculations is Johnson's Embedded-Atom Method (EAM) potential (Johnson 1988). The calculations characterize the dislocation structures and kinetics that mediate cavitation and the early stages of void growth, as well as the transient temperature fields attendant to the evolving dislocation structures. We also compute the evolution of macroscopic pressure with volumetric expansion and, in particular, characterize the cavitation pressure as a function of temperature and strain rate. These results are directly relevant to the formulation of models of nanovoid nucleation in metals subjected to shock conditions (Reina et al. 2011).

2 Finite-temperature non-equilibrium quasicontinuum

In this section we summarize the Kulkarni et al. (2008) extension of the theory of the quasicontinuum of Tadmor et al. (1996) (cf. Knap and Ortiz (2001) for the quasicontinuum implementation employed in the present work) to non-equilibrium systems at finite temperature. The extension relies on Jaynes' principle of maximum entropy (Jaynes 1957a,b) and on variational mean-field theory [cf., e.g., (Yeomans 1992) for a concise account] in order to construct effective thermodynamic potentials for systems of atoms away from equilibrium; and on the Yang et al. (2006) variational formulation of coupled transport problems in order to couple the non-equilibrium thermodynamic potentials to the heat equation. A fundamental difference with previous approaches based on free-energy minimization (cf., e.g., Sutton et al. 1992; Wang et al. 1992) is that the free energy formulated here pertains to systems away from equilibrium. In particular, the temperature field is not required to be uniform [cf., Sect. 2.1] but, contrariwise, it is allowed to take different values from atom to atom.

2.1 Maximum-entropy framework

We consider throughout a system consisting of N identical atoms. Thus, the *microscopic states* of the system are defined by the instantaneous atomic positions $\{\boldsymbol{q}\} = (\boldsymbol{q}_i)_{i=1}^N$ and momenta $\{\boldsymbol{p}\} = (\boldsymbol{p}_i)_{i=1}^N$. Suppose that the function $A(\{\boldsymbol{q}\}, \{\boldsymbol{p}\})$ represents a physical property of the system. Then, its expected or *macroscopic value* is given by the *phase average*

$$\langle A \rangle = \int A(\{\boldsymbol{q}\}, \{\boldsymbol{p}\}) p(\{\boldsymbol{q}\}, \{\boldsymbol{p}\})\, d\boldsymbol{q}\, d\boldsymbol{p}, \tag{1}$$

where $p(\{\boldsymbol{q}\}, \{\boldsymbol{p}\})$ is the *probability density function* of the system (cf., e.g., Hill 1962; Weiner 2002). In (1) and subsequently we write

$$d\boldsymbol{q}\, d\boldsymbol{p} = \left(\prod_{i=1}^{N}\prod_{j=1}^{3} dq_{ij}\right)\left(\prod_{i=1}^{N}\prod_{j=1}^{3} dp_{ij}\right) \tag{2}$$

for the element of volume in phase space.

In the present work, we assume that the statistics of the system obeys *Jaynes' principle of maximum entropy* (Jaynes 1957a,b). This principle postulates that the probability density function $p(\{\boldsymbol{q}\}, \{\boldsymbol{p}\})$ of the system maximizes the *information-theoretical entropy*

$$S[p] = -k_B \langle \log p \rangle, \tag{3}$$

subject to all the known constraints acting on the system. In (3) and subsequently, k_B denotes Bolzmann's constant. The maximum-entropy—or *max-ent* for short—probability density function is the *least biased* probability density function consistent with the constraints on the system. In order to allow for non-equilibrium processes we subject the system to *local constraints*, i.e., we presume that certain local or atom-wise averages are known. Specifically, we consider the following local constraints (Kulkarni et al. 2008):

i. The probability distribution satisfies the normalization condition

$$\langle 1 \rangle = 1 \tag{4}$$

ii. Each atom i moves about a mean or macroscopic position $\bar{\boldsymbol{q}}_i$ with a linear momentum that oscillates about a mean or macroscopic value $\bar{\boldsymbol{p}}_i$, i. e.,

$$\langle \boldsymbol{q}_i \rangle = \bar{\boldsymbol{q}}_i, \tag{5a}$$

$$\langle \boldsymbol{p}_i \rangle = \bar{\boldsymbol{p}}_i. \tag{5b}$$

iii. For each atom i, the variance $3\sigma_i$ of its momentum and the variance $3\tau_i$ of its atomic position are assumed known, i. e.,

$$\langle |\boldsymbol{p}_i - \bar{\boldsymbol{p}}_i|^2 \rangle = 3\sigma_i^2, \tag{6a}$$

$$\langle |\boldsymbol{q}_i - \bar{\boldsymbol{q}}_i|^2 \rangle = 3\tau_i^2. \tag{6b}$$

Introducing the local frequencies

$$\omega_i = \frac{\sigma_i}{m_i \tau_i}, \tag{7}$$

where m_i is the mass of atom i, and adding (5a) and (5b) we obtain the alternative local constraint

$$\langle \epsilon_i \rangle = \bar{\epsilon}_i, \tag{8}$$

where

$$\epsilon_i = \frac{1}{2m_i}\langle |\boldsymbol{p}_i - \bar{\boldsymbol{p}}_i|^2 \rangle + \frac{m_i \omega_i^2}{2}\langle |\boldsymbol{q}_i - \bar{\boldsymbol{q}}_i|^2 \rangle \tag{9}$$

may be regarded as a *meanfield or trial energy* of atom i, with the optimal *meanfield frequency* ω_i to be determined variationally, and

$$\bar{\epsilon}_i = \frac{3\sigma_i^2}{m_i} \tag{10}$$

its macroscopic average. By virtue of the preceding constraints, it follows that the *macroscopic state* of the system is defined by the data $\{\bar{\boldsymbol{q}}\} = (\bar{\boldsymbol{q}}_i)_{i=1}^N$, $\{\bar{\boldsymbol{p}}\} = (\bar{\boldsymbol{p}}_i)_{i=1}^N$, $\{\omega\} = (\omega_i)_{i=1}^N$ and $\{\bar{\epsilon}\} = (\bar{\epsilon}_i)_{i=1}^N$. It should be carefully noted that the macroscopic state of the system is presumed known solely for purposes of application of the maximum-entropy principle. In general, the determination of the time evolution of the macroscopic state requires the solution of an effective macroscopic problem, to be defined in the sequel.

According to the maximum-entropy principle, the probability distribution of the system follows as the critical point of the Lagrangian

$$L[p; \alpha, \{\beta\}] = -k_B \left(\langle \log p \rangle + \alpha \langle 1 \rangle + \{\beta\}^T \{\langle \epsilon \rangle\} \right), \tag{11}$$

where α and $\{\beta\} = (\beta_i)_{i=1}^N$ are Lagrange multipliers. Stationarity of $L[p; \alpha, \{\beta\}]$ demands that

$$p_0(\{q\}, \{p\}) = \frac{1}{Z_0} e^{-\{\beta\}^T \{\epsilon\}}, \tag{12}$$

where $Z_0 = e^{1+\alpha}$ is the *partition function*. Defining

$$k_B T_i \equiv \beta_i^{-1}, \tag{13}$$

by analogy of (12) to Gibbs' distribution T_i may be though of as the local temperature of atom i. Inserting (12) into the normalization condition Eq. (4) and evaluating the resulting Gaussian integrals, the partition function Z_0 is found to be

$$Z_0 = \prod_{i=1}^N Z_{0i}, \tag{14}$$

with

$$Z_{0i} = \left(\frac{2\pi}{\beta_i \omega_i} \right)^3. \tag{15}$$

Inserting Eq. (12) into (8) we additionally obtain

$$k_B T_i = \frac{1}{\beta_i} = \bar{\epsilon}_i, \tag{16}$$

which identifies $k_B T_i$ or, equivalently, β_i^{-1}, as the mean meanfield energy of atom i.

By virtue of the chosen macroscopic constraints, the resulting *max-ent* probability density function (12) and the partition function (14) depend parametrically on the *macroscopic state* of the lattice, namely, $(\{\bar{q}\}, \{\bar{p}\}, \{\beta\}, \{\omega\})$. As already noted, the evolution of the macroscopic state of the lattice remains to be determined as the solution of an effective macroscopic problem, to be defined. In particular, the formulation allows for *non-equilibrium* macroscopic states of the lattice, in which the deformation and thermal fields are non-uniform.

2.2 Variational meanfield free energy

Suppose that the Hamiltonian of the system is of the form

$$H(\{q\}, \{p\}) = \sum_{i=1}^N \frac{1}{2} \frac{|p_i|^2}{m_i} + V(\{q\}), \tag{17}$$

where $V(\{q\})$ is an interatomic potential. Introduce the *meanfield or trial Hamiltonian*

$$H_0(\{q\}, \{p\}) = \sum_{i=1}^N \epsilon_i, \tag{18}$$

with ϵ_i as in (9). We then define the *meanfield free energy* of the system as (Yeomans 1992)

$$F(\{\bar{q}\}, \{T\}) = \min_{\{\omega\}} \left\{ F_0 + \langle H - H_0 \rangle_0 \right\}, \tag{19}$$

where

$$F_0 = -\sum_{i=1}^N \frac{1}{\beta_i} \log \left(\frac{Z_{0i}}{h^3} \right), \tag{20}$$

and, here and subsequently, $\langle \cdot \rangle_0$ denotes phase-averaging with respect to the *max-ent* distribution p_0. In writing (20) we have introduced Planck's constant h for purposes of normalization. In particular, we note that the atomic frequencies $\{\omega\}$ are optimized variationally so as to yield the least possible meanfield free energy. The atomic frequencies resulting from the minimization in (19) approximate the actual potential energy of the system as best as possible in a variational meanfield sense (Yeomans 1992). A direct calculation gives

$$F_0 - \langle H_0 \rangle_0 = -\sum_{i=1}^N 3 k_B T_i \left[1 + \log \left(\frac{k_B T_i}{\hbar \omega_i} \right) \right], \tag{21}$$

where $\hbar = h/2\pi$ is the reduced Plank's constant. In addition we have

$$\langle H \rangle_0 = \sum_{i=1}^N \frac{3 k_B T_i}{2} + \langle V \rangle_0. \tag{22}$$

Combining (21) and (22) finally gives the meanfield free energy (19) as

$$F(\{\bar{q}\}, \{T\}) = \min_{\{\omega\}} \left\{ \langle V \rangle_0(\{\bar{q}\}, \{T\}, \{\omega\}) - \sum_{i=1}^N 3 k_B T_i \times \left[\frac{1}{2} + \log \left(\frac{k_B T_i}{\hbar \omega_i} \right) \right] \right\}, \tag{23}$$

As may be seen from this expression, the evaluation of the *effective interatomic potential* $\langle V \rangle_0$ requires the computation of a phase average of the interatomic potential V with respect to the *max-ent* probability distribution density p_0. The empirical interatomic potentials encountered in practice often have the additive structure

$$V(\{\boldsymbol{q}\}) = \sum_{i=1}^{N} V_i(\{\boldsymbol{q}\}), \tag{24}$$

where $V_i(\{\boldsymbol{q}\})$ is short-ranged and, consequently depends on $\boldsymbol{q}_i$ and on the positions of a small number of neighboring atoms. The effective interatomic potential then follows as

$$\langle V \rangle = \sum_{i=1}^{N} \langle V_i \rangle_0, \tag{25}$$

where the calculation of the local effective interatomic potential $\langle V_i \rangle_0$ takes places over a local phase subspace of small dimensionality. Since p_0 is Gaussian, the attendant local phase averages can conveniently be computed by numerical quadrature. In the present work, all phase averages with respect to p_0 are computed using third-order Gaussian-quadrature rules. Further details regarding the use of Gaussian quadrature for the computation of effective interatomic potentials may be found in Kulkarni et al. (2008).

2.3 Coupling to heat conduction

For a fixed temperature field $\{T\}$, the corresponding stable configurations $\bar{\boldsymbol{q}}^*(\{T\})$ of mechanical equilibrium of the system follow from the principle of minimum free energy, namely,

$$\bar{\boldsymbol{q}}^*(\{T\}) \in \text{argmin}\, F(\cdot, \{T\}), \tag{26}$$

where argmin $F(\cdot, \{T\})$ is the set of minimizers of $F(\{\bar{\boldsymbol{q}}\}, \{T\})$ with respect to $\{\bar{\boldsymbol{q}}\}$ at fixed $\{T\}$.

In order to determine the evolution of the temperature field, following Yang et al. (2006) we proceed to couple the equations (26) of mechanical equilibrium to a *discrete heat equation* expressed in entropy form, namely,

$$\dot{S}_i = \frac{\kappa}{T_i} \Delta \log\left(\frac{T_i}{T_0}\right), \tag{27}$$

where

$$S_i = -\frac{\partial F}{\partial T_i}(\{\bar{\boldsymbol{q}}\}, \{T\}) \tag{28}$$

is the *local entropy* of atom i, $\Delta \log T_i$ is a discrete Laplacian evaluated at atom i, T_0 is a reference absolute temperature and κ is an *empirical* thermal conductivity. In the calculations presented here, we define the local Laplacian simply by means of an auxiliary finite-element discretization based on a triangulation of the atomic positions (Kulkarni et al. 2008).

Equations (19), which determines the optimal atomic frequencies in the sense of variational meanfield theory, (26), which enforces the stable mechanical equilibrium of the system, and (27), which models heat conduction, constitute a closed set of coupled algebraic and ordinary differential equations that yield the time evolution of the mean atomic positions, $\{\bar{\boldsymbol{q}}(t)\}$, and temperature field, $\{T(t)\}$. This evolution can be further approximated in time by recourse to variational time discretization (Yang et al. 2006; Kulkarni et al. 2008).

2.4 Quasicontinuum reduction

The chief objective of the quasicontinuum method is to systematically coarsen an atomistic description by the judicious introduction of constraints (Tadmor et al. 1996; Miller and Tadmor 2002). These kinematic constraints are designed so as to preserve full atomistic resolution where required and to treat collectively large number of atoms in regions where the fields vary slowly on the scale of the lattice. The specific implementation of the quasicontinuum method employed in the calculations presented here is due to Knap and Ortiz (2001), as extended to non-equilibrium finite temperature by Kulkarni et al. (2008). Following Yang et al. (2006) we begin by noting that the governing equations (26) and the backward-Euler time discretization of (27) jointly follow as Euler-Lagrange equations of a coupled functional that depends on the mean atomic displacements $\{\bar{\boldsymbol{q}}\}$ and temperature field $\{T\}$. The desired degree-of-freedom reduction is then achieved by restricting the functional to subspaces spanned by finite-element shape functions based on the triangulation of a suitably chosen set of *representative atoms*. The selection of these atoms is effected adaptively so as to provide full atomistic resolution in the vicinity of lattice defects such as vacancies, dislocations and surface steps. Finally, the reduced Euler-Lagrange equations are evaluated using *summation rules*, akin to quadrature rules in finite elements (Knap and Ortiz 2001). Further details of implementation may be found in Knap and Ortiz (2001) and Kulkarni et al. (2008).

3 Void growth in copper at finite temperature

In this section we apply HotQC to the simulation of quasistatic void growth evolution in copper at finite temperature under triaxial expansion. The computational cell under consideration is 78 nm in size, encompasses 216^3 lattice cells for a total of 5.1×10^6 atoms, and is aligned with the cubic axes of the crystal. A spherical 2.0 nm void is introduced at the center of the cell. Figure 1a shows a close-up view of the initial mesh used in the calculations, which provides full atomistic resolution within a $8a \times 8a \times 8a$ region. Away from this fully-atomistic region, the initial triangulation is gradually coarsened with the result that the initial computational mesh contains 900 representative atoms only. The mesh is subsequently adapted so as to ensure that all dislocations are embedded within a sizeable atomistic region at all times (Knap and Ortiz 2001), cf. Fig. 1b. The computational cell is deformed to 30% triaxial or volumetric engineering strain by means of displacement control at its outer boundary. In addition, the computational domain is assumed to be thermally insulated. The deformations are imparted at 0.1% increments and the corresponding successive equilibrium configurations of the crystal are calculated by means of the Polak-Ribiere algorithm (Vetterling et al. 2002) or, if convergence is too slow, dynamic relaxation (Oakley and Knight 1995; Oakley et al. 1995; Underwood 1983). The interatomic potential used in the calculations is Johnson's Embedded-Atom Method (EAM) potential (Johnson 1988).

In order to facilitate verification against previous work, we begin by deforming the crystal at high strain rates typical of molecular dynamics calculations. Figure 2 shows the computed pressure versus volumetric engineering strain ε_v for four different initial simulation temperatures, ranging from 150 to 600 K, and a strain rate $\dot{\varepsilon} = 2.5 \times 10^{11}\ \mathrm{s}^{-1}$.

As may be seen from Fig. 3, at a strain rate of $\dot{\epsilon} = 10^{10}\ \mathrm{s}^{-1}$, HotQC predicts a pressure-volume curve that lies somewhat below the curve obtained from molecular dynamics (**?**). A number of factors, including differences in the interatomic potential, may account for this discrepancy. Since the HotQC calculations presented here are quasistatic, which may also partly account for the discrepancy, it is instructive to estimate the range of strain rates for which microinertia, which is accounted for in the molecular dynamics

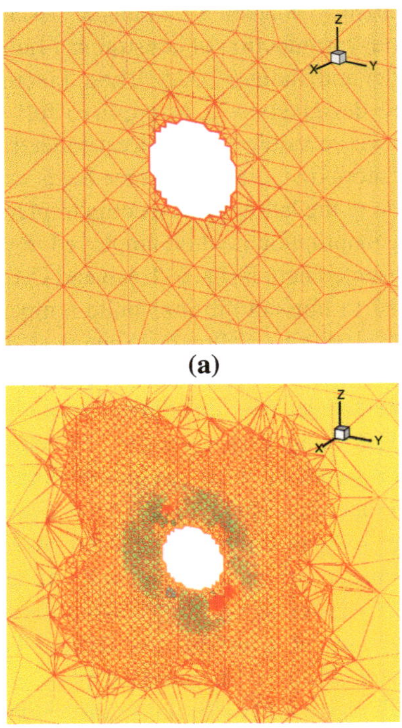

Fig. 1 Close-ups of finite element mesh used in the QC reduction. **a** Initial mesh, including a fully atomistic region around the void. **b** Typical final mesh, showing mesh refinement resulting in an extended fully-atomistic region that encompasses all the dislocations emitted by the void

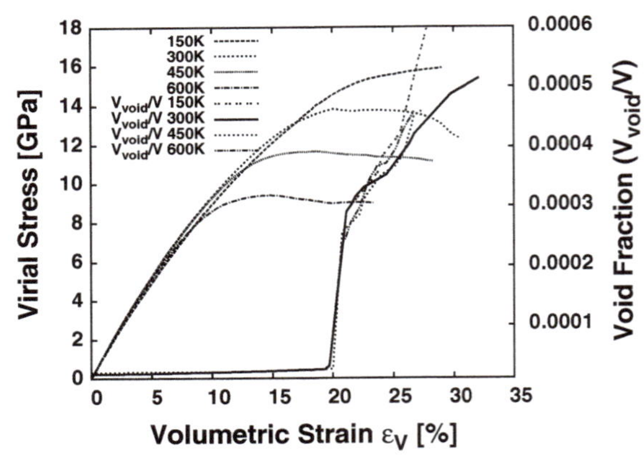

Fig. 2 Computed virial stress (pressure) and normalized void volume expansion versus volumetric engineering strain ε_v of the sample. The triaxial expansion is carried out at a strain rate of $2.5 \times 10^{11}\ \mathrm{s}^{-1}$ and different initial temperatures ranging from 150 to 600 K. The figure also plots the evolution of the void fraction at the same initial temperatures

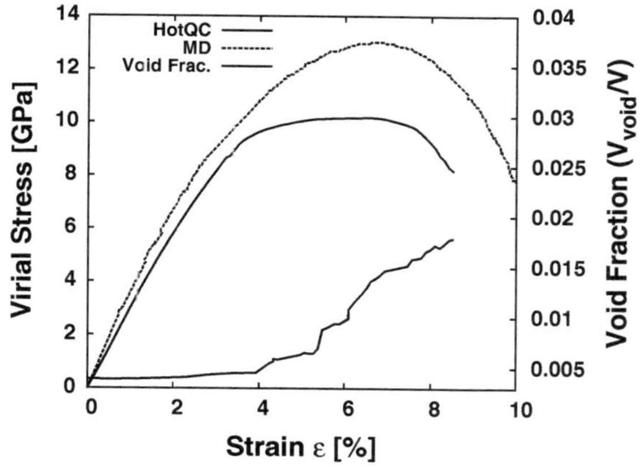

Fig. 3 Comparison of virial stress (pressure) and normalized void volume expansion versus engineering strain ε predicted by HotQC and obtained from molecular dynamics (?) for the same simulation sample (72^3 lattice cells and a void of radius 2.2 nm). The triaxial expansion is carried out at a strain rate of 10^{10} s^{-1}

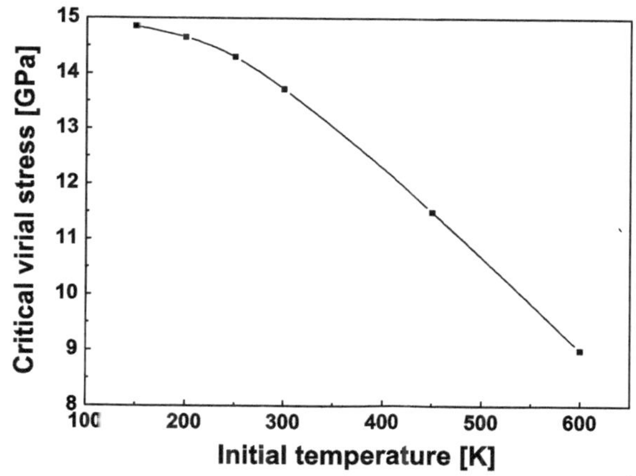

Fig. 4 Computed critical virial stress (pressure) versus temperature in the range of 150–600 K at a strain rate of 2.5×10^{11} s^{-1}

calculations, may be expected to play a significant role during void growth. Ortiz and Molinari (1992) analyzed the dynamic expansion of a spherical void in a rigid-plastic material and determined that inertia effects become important when the void size a is larger than an intrinsic characteristic length $l_c = C/\dot{\epsilon}$, where $\dot{\epsilon}$ is a representative strain rate, $C = \sqrt{\sigma/\rho}$, σ is a representative yield stress and ρ is the mass density. For copper, taking $\rho = 8.94$ Kg/dm^3, $\sigma = 200$ MPa, and a void radius $a = 2.2$ nm, the critical strain rate for which $a = l_c$ follows as $\dot{\epsilon}_c = 6.9 \times 10^{10}$ s^{-1}. Thus, for small voids inertia effects may be expected to be important only for exceedingly large strain rates. This finding is consistent with continuum analysis of the effect of inertia on void growth in plastic materials (Ortiz and Molinari 1992; Weinberg et al. 2006), which suggest that inertia is negligible for small enough voids. The strain rate of $\dot{\epsilon} = 2.5 \times 10^{11}$ s^{-1} used in the calculations shown in Fig. 2, and representative of previous molecular dynamics calculations (?) is, in fact, supercritical, which is likely to partly account for the discrepancy between the HotQC and molecular dynamics results. It should be noted that, even though inertia is not taken into account in the HotQC calculations presented here, the model still contains an intrinsic time scale set by heat conduction. This intrinsic time scale discriminates between slow processes, which tend to be isothermal, and fast processes, which tend to be adiabatic. It should also be noted that HotQC is amenable to a straightforward dynamical implementation, e.g., by recourse to variational time integrators (Radovitzky and Ortiz 1999).

Figure 2 also shows the evolution of the void volume fraction V_{void}/V for the same range of initial temperatures. As expected, following an initial elastic regime, the response of the crystal softens causing the pressure to go through a maximum. Upon the attainment of a critical strain, of the order of 20%, dislocations are emitted from the void and the rate of growth of the void increases precipitously. This sharp transition is clearly discernible from the figure and can be identified with the onset of cavitation, and the corresponding pressure with the critical pressure for cavitation. The dependence of the cavitation pressure on temperature is shown in Fig. 4. As may be seen from the figure, the cavitation pressure is relatively temperature-insensitive at low temperatures and decreases markedly beyond a transition temperature of the order of 250 K.

As discussed in Sect. 2, HotQC accounts for non-equilibrium conditions, including heat conduction, which permits tracking the evolution of the temperature field induced by dislocation emission. Figure 5 shows the computed temperature field in the vicinity of the void at 22% prescribed volumetric strain and 2.5×10^{11} s^{-1} strain rate. Remarkably, the crystal heats up locally around the void by over 100 K as a result of dislocation emission. However, the rate of heating is by no means uniform or constant in time. Thus, Fig. 6 shows the time-evolution of the temperature at three points within the sample: one point on the surface of

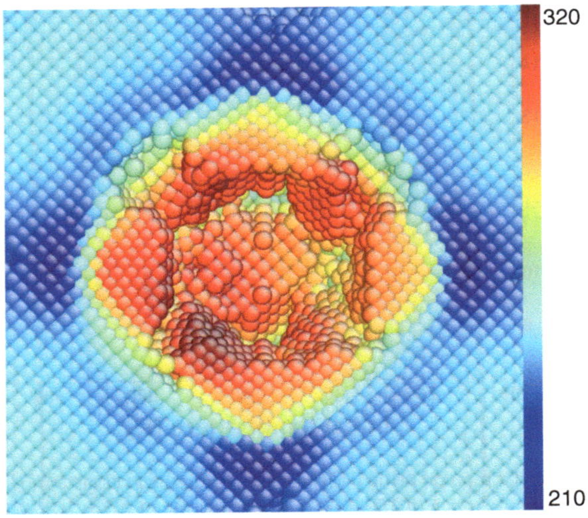

Fig. 5 Temperature field in the vicinity of the void at 22% prescribed volumetric strain

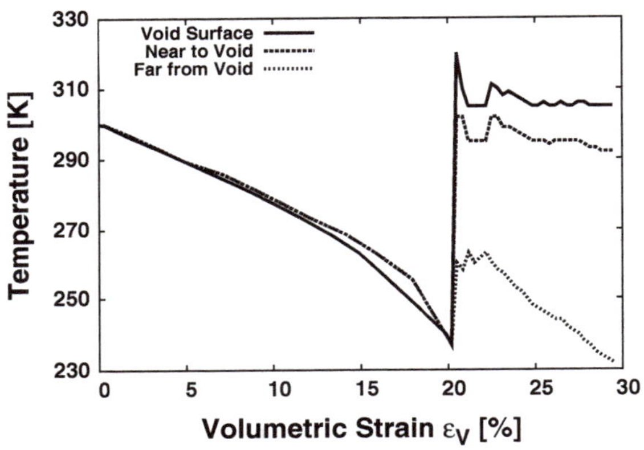

Fig. 6 Temperature evolution during void growth at different positions within the sample. **a** Void surface. **b** Position near the void surface. **c** Position far from the void surface

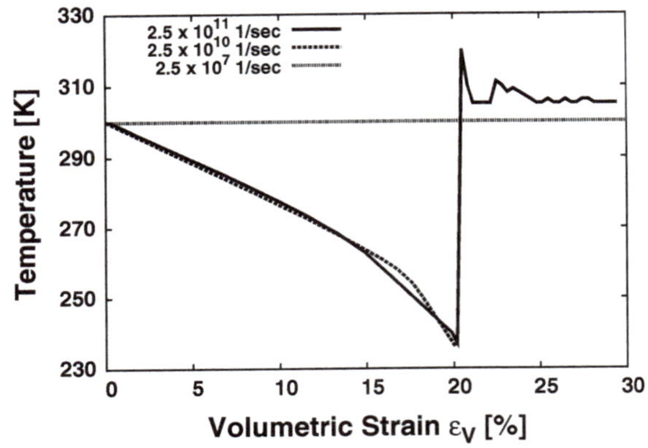

Fig. 7 Temperature versus volumetric deformation ε_v at strain rates 2.5×10^{11}, 2.5×10^{10} and 2.5×10^{7} s^{-1}

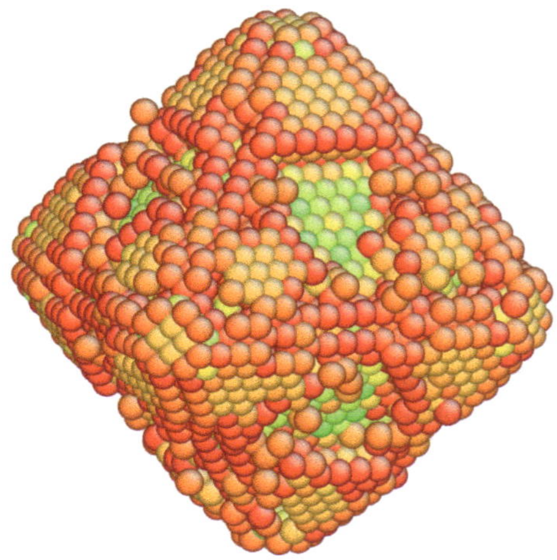

Fig. 8 Triaxial expansion of void at an initial temperature of 300 K and a strain rate of $2.5 \times 10^{11} s^{-1}$. Octahedral shape of the atoms encompassing the surface of the void at a strain of ~14.4%

the void, a second point near the void and a third point far from the void. As expected, at large strains the temperature decreases monotonically with distance to the void. Remarkably, prior to the attainment of the critical volumetric strain for dislocation emission, the crystal cools down due to the thermoelastic effect. By contrast, following the critical volumetric strain, dislocation emission causes a rapid local heating of the crystal. This heating is highly inhomogeneous and concentrated around the void, which in turn sets up a temperature gradient that results in the conduction of heat away from the void. Figure 7 depicts the evolution of the temperature field near the surface of the void for three strain rate spanning the range of 2.5×10^{11} to 2.5×10^{7} s^{-1}. As may be seen from the figure, the temperature evolution is ostensibly rate-independent.

A detailed analysis of the dislocation structures that mediate void cavitation is beyond the scope of this paper. Here we limit ourselves to presenting evidence that, indeed, cavitation is accompanied by the emission of dislocations from the void. Prior to cavitation, the surface of the void reconstructs and adopts an octahedral shape, cf. Fig. 8. This finding is in keeping with observations of void evolution in FCC metals during spall experiments (Stevens et al. 1972). Figure 9 collects snapshots of dislocations emitted from

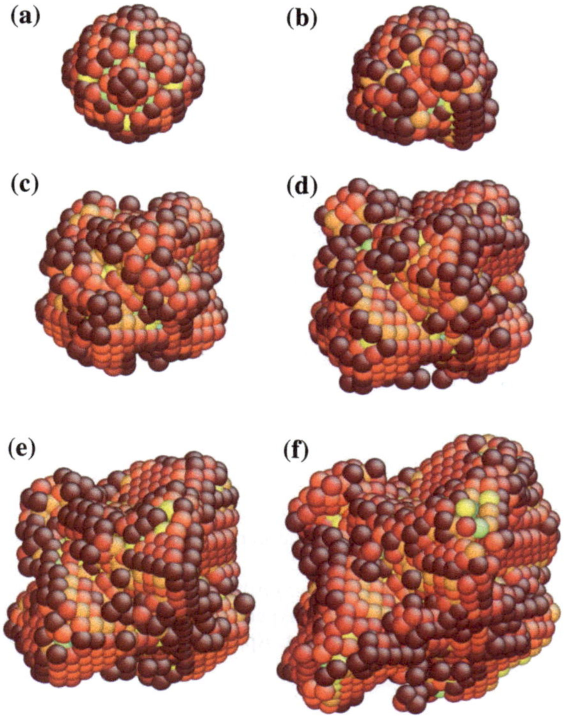

Fig. 9 Snapshots of the atoms in the vicinity of the void at different strains during triaxial expansion. **a** $\varepsilon_v = 19.7\%$, **b** $\varepsilon_v = 20\%$, **c** $\varepsilon_v = 20.7\%$, **d** $\varepsilon_v = 21.5\%$, **e** $\varepsilon_v = 23\%$, and **f** $\varepsilon_v = 25\%$

the void at cavitation. The dislocations are identified by means of the centrosymmetry deviation parameter (Kelchner et al. 1998). Remarkably, dislocations are entirely absent from the crystal right up to cavitation, and are emitted abruptly and profusely over an exceedingly narrow volumetric strain interval following cavitation. The suddenness and swiftness of the process is consistent with the bifurcation interpretation of the cavitation of voids in solids (Ball 1982).

4 Summary and concluding remarks

We have applied the HotQC method of Kulkarni et al. (2008) to the study of quasistatic void growth in copper single crystals at finite temperature under triaxial expansion. The computational model consists of a spherical 2.0 nm void at the center 78 nm cubic cell aligned with the cubic lattice directions. The computational cell is strained to 30% deformation at initial temperatures and nominal strain rates ranging from 150 to 600 K and from 2.5×10^5 to 2.5×10^{11} s^{-1}, respectively. The interatomic potential used in the calculations is Johnson's Embedded-Atom Method (EAM) potential Johnson (1988). The computed pressure versus volumetric strain is in close agreement with that obtained by (?Bringa et al. 2010) using molecular dynamics, which suggests that inertia effects are not dominant for the void size and conditions considered. We find that, upon the attainment of a critical or cavitation strain of the order of 20%, dislocations are abruptly and profusely emitted from the void and the rate of growth of the void increases precipitously. The cavitation pressure is found to be relatively temperature-insensitive at low temperatures and decreases markedly beyond a transition temperature of the order of 250 K. Remarkably, prior to cavitation, the crystal cools down due to the thermoelastic effect. By contrast, following cavitation dislocation emission causes rapid local heating in the vicinity of the void, which in turn sets up a temperature gradient and results in the conduction of heat away from the void. It bears emphasis that all results presented in this paper correspond to Johnson's EAM potential for copper. A question of considerable interest concerns the sensitivity of cavitation in FCC metals to the choice of interatomic potential.

In closing, we remark on the ability of HotQC to track non-equilibrium thermal effects, including heat conduction, at the atomic level. This ability comes at the expense of introducing an additional empirical constant, namely, the thermal conductivity κ cf. (27). However, this level of empiricism is not greater than—indeed, it is commensurate with—that involved in formulating interatomic potentials, which themselves are also empirical in nature. The great advantage of HotQC is that it entirely bypasses the need to track the thermal fluctuations of the atoms in detail, a requirement that often besets molecular dynamics. In particular, the time steps required for the numerical solution of the problem are set by the diffusional time scale, which is much greater than the time scale of the atomic vibrations. In this manner, HotQC enables the tracking of diffusive processes over very long periods of time, while simultaneously building atomistic realism into the model.

Acknowledgements We gratefully acknowledge the support of the Ministerio de Ciencia e Innovación of Spain (DPI2009-14305-C02-01/02) and the support of the Consejería de Innovación of Junta de Andalucía (P09-TEP-4493). Support for this study was also provided by the Department of Energy National Nuclear Security Administration under Award Number DE-FC52-08NA28613 through Caltech's ASC/PSAAP Center for the Predictive Modeling and Simulation of High Energy Density Dynamic Response of Materials.

References

Ahn DC, Sofronis P, Kumar M, Belak J, Minich R (2007) Void growth by dislocation-loop emission. J Appl Phys 101: 063,514

Aoki S, Kishimoto K, Takeya A, Sakata M (1984) Effects of microvoids on crack blunting and initiation in ductile materials. Int J Fract 24:267–278

Argon AS, Im J, Safoglu R (1975) Cavity formation from inclusions in ductile fracture. Metall Trans A 6A:825–837

Ball JM (1982) Discontinuous equilibrium solutions and cavitation in nonlinear elasticity. Philos Trans R Soc Lond Ser A Math Phys Sci 306:557–611

Bauer RW, Wilsdorf HGF (1973) Void initiation in ductile fracture. Scripta Metall 7:1213–1220

Belak J (1998) On the nucleation and growth of voids at high strain-rates. J Comput Aided Mater Des 5:193–206

Belak J (2002) Multi-scale applications to high strain-rate dynamic fracture. J Comput Aided Mater Des 9:165–172

Belak J (2005) Incipient spallation fracture in light metals from 3D X-ray tomography, 2D microscopy, and molecular dynamic simulations. In: 11th international conference on fracture

Bringa E, Traiviratana S, Meyers MA (2010) Void initiation in fcc metals: effect of loading orientation and nanocrystalline effects. Acta Mater 58(13):4458–4477

Dávila LP, Erhart P, Bringa EM, Meyers MA, Lubarda VA, Schneider MS, Becker R, Kumar M (2005) Atomistic modeling of shock-induced void collapse in copper. Appl Phys Lett 86:161,902

Dupuy LM, Tadmor EB, Miller RE, Phillips R (2005) Finite-temperature quasicontinuum: molecular dynamics without all the atoms. Phys Rev Lett 95: 060202-1–060202-4

Garrison WM, Moody NR (1987) Ductile fracture. J Phys Chem Solids 48:1035–1074

Hancock JW, Mackenzie AC (1976) On the mechanisms of ductile failure in high-strength steels subjected to multi-axial stress-states. J Mech Phys Solids 24:147–169

Hill T (1962) Introduction to statistical thermodynamics. Addison Wesley, Boston

Jaynes ET (1957a) Information theory and statistical mechanics i. Phys Rev Ser II 106(4):620–630

Jaynes ET (1957b) Information theory and statistical mechanics ii. Phys Rev Ser II 108(2):171–190

Johnson J (1981) Dynamic fracture and spallation in ductile solids. J Appl Phys 52(4):2812–2825

Johnson RA (1988) Analytic nearest-neighbor model for fcc metals. Phys Rev B 37:3924–3931

Kelchner CL, Plimpton SJ, Hamilton JC (1998) Dislocation nucleation and defect structure during surface nanoindentation. Phys Rev B 58:11085–11088

Knap J, Ortiz M (2001) An analysis of the quasicontinuum method. J Mech Phys Solids 49:1899–1923

Kulkarni Y, Knap J, Ortiz M (2008) A variational approach to coarse graining of equilibrium and non-equilibrium atomistic description at finite temperature. J Mech Phys Solids 56:1417–1449

Lubarda VA, Schneider MS, Kalantar DH, Regminton BA, Meyers MA (2004) Void growth by dislocation emission. Acta Mater 52:1397–1408

Marian J, Knap J, Ortiz M (2004) Nanovoid cavitation by dislocation emission in aluminum. Phys Rev Lett 93:165503-1–165503-4

Marian J, Knap J, Ortiz M (2005) Nanovoid deformation in aluminum under simple shear. Acta Mater 53:2893–2900

Marian J, Venturini G, Hansen BL, Knap J, Ortiz M, Campbell GH (2010) Finite-temperature extension of the quasicontinuum method using langevin dynamics: entropy losses and analysis of errors. Model Simul Mater Sci Eng 18:015, 003

Meyers MA, Aimone CT (1983) Dynamic fracture (spalling) of metals. Prog Mater Sci 28(1):1–96

Meyers MA, Traiviratana S, Lubarda VA, Benson DJ, Bringa EM (2009) The role of dislocations in the growth of nanosized voids in ductile failure of metals. JOM J Miner Met Mater Soc 61:35–41

Miller RE, Tadmor EB (2002) The quasicontinuum method: overview, applications and current directions. J Comput Aided Mater Des 9:203–239

Oakley DR, Knight NF (1995) Adaptive dynamic relaxation algorithm for non-linear hyperelastic structures part i. formulation. Comput Methods Appl Mech Eng 126(1–2):67–89

Oakley DR, Knight NF, Warner DD (1995) Adaptive dynamic relaxation algorithm for non-linear hyperelastic structures part iii. parallel implementation. Comput Methods Appl Mech Eng 126(1–2):111–129

Ortiz M, Molinari A (1992) Effect of strain-hardening and rate sensitivity on the dynamic growth of a void in a plastic material. ASME J Appl Mech 59(1):48–53

Puttick KE (1959) Ductile fracture in metals. Philos Mag 4:964–969

Radovitzky R, Ortiz M (1999) Error estimation and adaptive meshing in strongly nonlinear dynamic problems. Comput Methods Appl Mech Eng 172:203–240

Reina C, Marian J, Ortiz M (2011) Nanovoid nucleation by vacancy aggregation and vacancy-cluster coarsening in high-purity metallic single crystals. Phys Rev B 84(10):104117-1–104117-12

Rudd RE, Belak JF (2002) Void nucleation and associated plasticity in dynamic fracture of polycrystalline copper: an atomistic simulation. Comput Mater Sci 24:148–153

Seppälä ET, Belak J, Rudd RE (2004a) Effect of stress triaxiality on void growth in dynamic fracture of metals: a molecular dynamics study. Phys Rev B 69:134101-1–134101-19

Seppälä ET, Belak J, Rudd RE (2004b) Onset of void coalescence during dynamic fracture of ductile metals. Shock Compress Condens Matter 93:491–494

Stevens A, Davison L, Warren W (1972) Spall fracture in aluminum monocrystals—dislocation-dynamics approach. J Appl Phys 43(12):4922–4927

Strachan A, Cagin T, III WAG (2001) Critical behavior in spallation failure of metals. Phys Rev B 63:060,103

Sutton A, Mulheran P, Stoneham A (1992) Direct free energy minimization methods: application to grain boundaries. Philos Trans Phys Sci Eng 341(1661):233–245

Tadmor EB, Ortiz M, Phillips R (1996) Quasicontinuum analysis of defects in solids. Philos Mag 73:1529–1563

Tang Z, Zhao H, Li G, Aluru NR (2006) Finite-temperature quasicontinuum method for multiscale analysis of silicon nanostructures. Phys Rev B 74(6):064110-1–064110-16

Traiviratana S, Bringa EM, Benson DJ, Meyers MA (2008) Void growth in metals: atomistic calculations. Acta Mater 56:3874–3886

Tvergaard V (1990) Material failure by void growth to coalescence. Adv Appl Mech 27:83–151

Underwood P (1983) Dynamic relaxation. In: Hughes TJR, Belytschko TB (eds) Computational methods for transient analysis. Elsevier, Netherlands, pp. 245–265

Vetterling W, Press W, Flannery B, Teukolsky S, Press WH (2002) Numerical recipes example book C++. Cambridge University Press, Cambridge

Wang H, Najafabadi R, Srolovitz D (1992) (100) Surface segregation in Cu–Ni alloys. Phys Rev B 45(20):12028–12042

Weinberg K, Mota A, Ortiz M (2006) A variational constitutive model for porous metal plasticity. Comput Mech 37:142–152

Weiner J (2002) Statistical mechanics of elasticity. Dover Publications, Mineola

Yang Q, Stainier L, Ortiz M (2006) A variational formulation of the coupled thermo-mechanical boundary-value problem for general dissipative solids. J Mech Phys Solids 54(2):401–424

Yeomans JM (1992) Statistical mechanics of phase transitions. Oxford University Press, Oxford

Int J Fract (2012) 174:87–94
DOI 10.1007/s10704-011-9656-0

ORIGINAL PAPER

Coupled phase transformations and plasticity as a field theory of deformation incompatibility

Amit Acharya · Claude Fressengeas

Received: 31 August 2011 / Accepted: 22 November 2011 / Published online: 13 December 2011

Abstract The duality between terminating discontinuities of fields and the incompatibilities of their gradients is used to define a coupled dynamics of the discontinuities of the elastic displacement field and its gradient. The theory goes beyond standard translational and rotational Volterra defects (dislocations and disclinations) by introducing and physically grounding the concept of *generalized disclinations* in solids without a fundamental rotational kinematic degree of freedom (e.g. directors). All considered incompatibilities have the geometric meaning of a density of lines carrying appropriate topological charge, and a conservation argument provides for natural physical laws for their dynamics. Thermodynamic guidance provides the driving forces conjugate to the kinematic objects characterizing the defect motions, as well as admissible constitutive relations for stress and couple stress. We show that even though higher-order kinematic objects are involved in the specific free energy, couple stresses may not be required in the mechanical description in particular cases. The resulting models are capable of addressing the evolution of defect microstructures under stress with the intent of understanding dislocation plasticity in the presence of phase transformation and grain boundary dynamics.

Keywords Phase transformation · Plasticity · Defects

A. Acharya (✉)
Civil & Environmental Engineering, Carnegie Mellon University, Pittsburgh, PA 15213, USA
e-mail: acharyaamit@cmu.edu

C. Fressengeas
Laboratoire d'Etude des Microstructures et de Mécanique des Matériaux (LEM3), Université Paul Verlaine-Metz/CNRS, Ile du Saulcy, 57045 Metz Cedex, France
e-mail: claude.fressengeas@univ-metz.fr

1 Introduction

The dynamic response of solids to mechanical and thermal loading involves material-physics phenomena that are manifested over multiple length and time scales. The material responds to load through atomic level processes involving crystal defects that mediate plasticity, phase transformations and damage. These processes, however, generally occur in materials whose intrinsic microstructure affects bulk properties like strength, ductility, and fracture toughness. Although relevant dimensions and time span for the nucleation of the individual defects can be Å and picoseconds, their collective interaction, evolution and relaxation occur at larger spatial and time scales (μm and μs). The objective of the program on which this paper is based is the development of a kinematically rigorous continuum mechanics model to serve as the basis for a simulation framework; the goal is to predict deformation-induced microstructure evolution and its effects on macroscopic properties in materials that undergo coupled plasticity and phase transformation. The kinematic ingredients of the model naturally include structural defects representative of grain boundaries. Such inelastic deformation is complex because of the essentially collective character of the dynamical behavior of structural defects over engineering time and length scales; nevertheless, understanding the underlying dynamics of defects is essential for predicting flow and failure at meso- and macroscopic scales. Due to space limitations, in this paper we outline only the small deformation kinematics and thermodynamics of our model.

This paper may be considered as an extension of the pioneering work of deWit (1970) on dislocation-disclination statics that was generalized by Fressengeas et al. (2011) to account for disclination-dislocation dynamics; our work describes the dynamics of a more general class of defects than dislocations and disclinations. It also has broad philosophical similarities with the work of Kleinert (2008). Here, the whole elastic distortion 'gradient' can be incompatible in order to represent terminating/kinking elastic distortion discontinuities as may arise, e.g., in a martensitic needle or a facet of a phase-transforming inclusion. Through the connection to the kinematics of phase transformations, our work provides an unambiguous physical basis to the notion of generalized disclinations (*g-disclinations*) even in solids without any director degrees of freedom. It is also motivated by that of Abeyaratne and Knowles (1990) on dynamic phase transformations that allows for full freedom in specifying kinetics of phase front motion, and generalizes it to account for geometric singularities in the surfaces of discontinuity. In this last context, there appear to be connections between our model and the fundamental work of Simha and Bhattacharya (1998), and through that, to the configurational force framework of Gurtin (2000). However, we believe that the details of generating approximate solutions to our theory based on nonlinear partial differential equations (PDE), while non-trivial, are going to be much simpler[1] than the complications involved in the configurational force setup that, to our knowledge, has not been solved in full 3-d generality as yet. As we understand, these difficulties stem from applying boundary conditions for the field equations on the (solution-dependent) moving boundaries of the cylinders representing the cores of the phase-transformation singularities. A primary future goal of our program is to make contact with the Topological Model put forward by Hirth and Pond (2011) in the modeling of displacive phase transformations.

Phase field methodology has been used to study phase transformations and plasticity (Levitas et al. 2009), but the governing dynamical principle there is of finding local minima of energy. In contrast, the dynamics we propose is based on geometrically rigorous conservation laws for line-defect densities carrying appropriate topological charge (much like vortices in fluids and superconductivity) and we have shown, in simple but exact realizations of sub-parts of our model, that the class of equilibria of such conservation laws contains that of phase field models (with the same energy) but is strictly larger, while in dynamic behavior it allows defect motion as wave propagation that simply cannot be predicted by phase field models (Acharya et al. 2010; Acharya and Tartar 2011). Also, generalizing the standard formalism for describing stress (arising from elastic straining) in phase field methodology to the finite deformation setting to incorporate elastic response developed from atomistic considerations appears to be an as yet unsettled matter.

We use more-or-less standard notation. All tensor indices are written with respect to the basis $e_i, i = 1$ to 3 of a rectangular Cartesian coordinate system. The symbol e_{ijk} represents the components of the alternating tensor X. Vertical arrays of two or three dots represent contraction of the respective number of 'adjacent' indices on two immediately neighboring symbols (in standard fashion). Also,

$$\left.\begin{aligned}(curl\ A)_{ij\text{ or }ikj} &= e_{jrs}\partial_r A_{is\text{ or }iks}\\ (div\ A)_{i\text{ or }ik} &= \partial_r A_{ir\text{ or }ikr}\\ (A \times V)_{ij\text{ or }ikj} &= e_{jrs}V_s A_{ir\text{ or }ikr}\end{aligned}\right\}$$

for A a 2nd or 3rd order tensor

2 The physical idea

The fundamental theoretical concern is the modeling of the dynamics of the relevant defects that may be viewed as *discontinuities* in the elastic displacement, rotation and distortion fields (slip, grain and phase boundaries, respectively) across surfaces, terminating in line *singularities* (dislocation, standard disclination and *generalized disclination*, respectively) as sketched in Fig. 1.

[1] Indeed, Fressengeas et al. (2011) have already developed and demonstrated a finite-element implementation of the pure disclination-dislocation dynamics case. An attractive feature is the coherence of the algorithm with that of the pure field dislocation dynamics case first introduced in Roy and Acharya (2005).

It is well known that when defects are viewed at sufficiently small scales there are no discontinuities and singularities but only appropriately localized smooth fields. Moreover, for the purpose of understanding the nonlinear dynamics of such defects and their collective effects, in particular in the context of nonlinear PDEs, such smooth localized representations of the physical reality of defects are essential from the point of view of mathematical analysis and numerical computation. There is a significant duality between the terminating curves of discontinuities of fields on 2-d surfaces and the *smooth incompatibilities* of the 'gradients' of such fields. The essential idea is that the geometrically defined incompatibilities model defect curves. Each (often closed) curve separates two disjoint pieces of a 2-d surface. The field in question is discontinuous by different amounts across the two 2-d pieces, and the advance of the curve represents the spreading of one piece at the expense of the other. The incompatibilities have the rigorous geometric meaning of a density of lines carrying appropriate topological charge, so that a tautological (and therefore, undeniable at the kinematic level) conservation argument provides most of the structural elements of a natural physical law for the dynamics of the field.[2] We exploit this feature to define coupled PDE-dynamics of the discontinuities of the elastic displacement field and its gradient. When restricted to the elastic displacement, a nonlinear theory for the dynamics of dislocation and slip results. When applied to the elastic distortion, the theory provides for the dynamics of *g-disclinations*, going beyond the Volterra construct by including elastic strain discontinuities, and allows dealing with phase and grain boundaries. Unlike the Volterra concept, our continuously distributed approach seamlessly models single defects up to a continuously distributed field in the entire body, as might be encountered in the transition of a solid to a liquid-like state under strong shock loading. These models are capable of addressing phase transformation and dislocation motion-related microstructure evolution under stress. While being based on PDEs in the set of variables involved, they are non-local models in space and time in the standard variables of conventional continuum mechanics, thus providing geometrically rigorous, physics-based, nonlocal generalization of the latter, and leading to well-set models of post-initiation defect mechanics.

[2] This is in contrast to dynamics based on the local "principle of virtual power" whose validity has recently been questioned (Fosdick 2011).

3 Motivating kinematical notions

As is well-known, a classical, singular dislocation line in a simply-connected body has a smooth elastic (1−)distortion (strain + rotation) field in the non simply-connected, punctured domain represented by the region excluding its core, even though the elastic displacement field it is considered to arise from has a discontinuity across a (non-unique) 2-d smooth slip surface in the same region. The geometry of any such 2-d surface is arbitrary, except that it terminates along the curve representing the core of the dislocation. Additionally, the important topological fact about such a field is that a line integral of the elastic distortion along any curve encircling the core is non-vanishing and constant over all such curves. This constant represents the strength of the dislocation, i.e., the *Burgers vector*. Thus, as concerns a classical dislocation, it is only its elastic distortion field that one can precisely discuss, and not the displacement field. The continuously distributed dislocations setting eases this classical description by considering smooth elastic distortion fields on simply connected domains in which they are pointwise irrotational outside the core cylinder (possibly of non-zero volume), and their curl equals the corresponding *Nye dislocation density* tensor fields inside the cylinder. The distribution of the elastic distortion and the dislocation density tensor fields on the body are set up such that the Burgers vector of the dislocation being modeled is obtained on integration of the fields along appropriate surface patches and/or closed curves.

Once understood, this fact enables the definition of a phase boundary, i.e., a 2-d surface of elastic distortion discontinuity (which includes the case of a grain boundary), terminating on a g-disclination curve. In this case, we assume the continuous elastic 2-distortion field (i.e., the incompatible elastic distortion 'gradient' field) to be irrotational outside the g-disclination core, and the g-disclination strength to be the second-order tensor obtained by integrating the 2-distortion field along a closed curve encircling the core. In the pure disclination case, this second-order tensor is a skew-tensor whose axial vector is called the *Frank vector* of the disclination. In the non-singular, continuously distributed

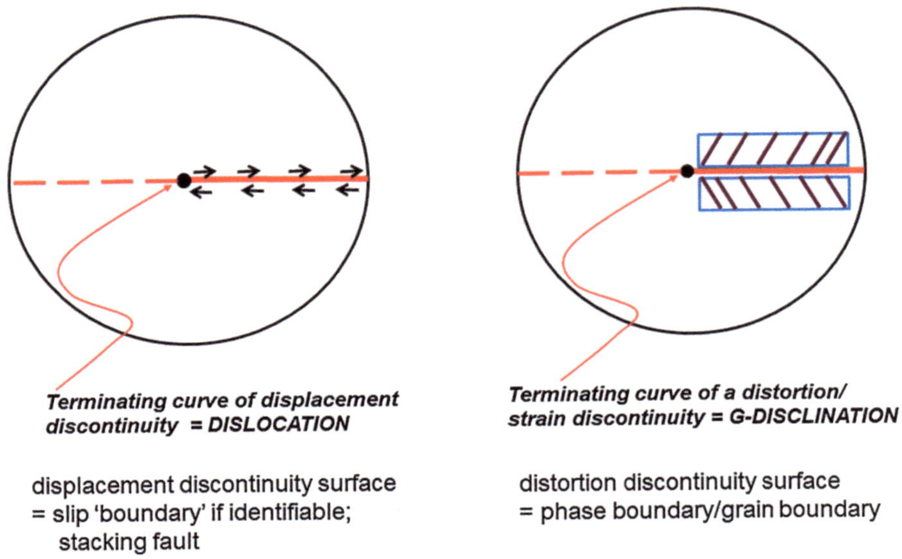

Fig. 1 Discontinuity of a discontinuity; cross-section view of straight defect lines terminating displacement/distortion discontinuities

setting, one way of setting up the 3rd order, g-disclination density field for a straight g-disclination would be to assign the tensor product of the 2nd order, strength tensor and the core line direction divided by the core cross-sectional area as a constant distribution within the core cylinder and zero outside it. Considered from the dual viewpoint of a discontinuous elastic distortion field with discontinuity terminating at a core singularity, the strength tensor is simply the jump in the elastic distortion across the surface of discontinuity.

It is important to note explicitly that, in the continuously distributed setting, the elastic distortions are no longer gradients but have incompatible parts that may be defined through a Helmholtz decomposition. We would now like to set up a dynamical theory of such dislocation and g-disclination defect curves and (meta)slipped regions, taking into account forces, moments and dissipation.

4 Mechanical structure

Let T be the (generally unsymmetric) stress tensor, Λ the couple stress tensor, X the alternating tensor, u the material displacement vector, v the material velocity vector, D and Ω the symmetric and skew-symmetric parts of the velocity gradient, and $\omega = -(1/2)\, X : \Omega = curl\, v$ the vorticity vector. Let U^e be the elastic 1-distortion tensor and G^e the elastic 2-distortion tensor (in gradient elasticity these would be $grad\, u$ and $gradgrad\, u$, respectively). We define

$$G^e - grad\, U^e =: P^s \tag{1}$$

$$-G^e : X =: \alpha^s \tag{2}$$

$$curl\, G^e = curl\, P^s = \Pi^s, \tag{3}$$

where α^s is the dislocation density (2nd-order) tensor and Π^s is the g-disclination density (3rd-order) tensor. Both the dislocation density and the g-disclination density tensors are 'vector(tensor)-valued axial vector' fields representing third (fourth) order tensor fields that are skew in their last two indices. Thus, they form densities that can be integrated over area elements. Physically, such fields represent densities of lines carrying appropriate tensorial attributes.

It is helpful for physical interpretation to invoke a Stokes–Helmholtz-like orthogonal decomposition of the field P^s into compatible and incompatible parts:

$$\left.\begin{array}{l} P^s = P^{s\perp} + grad\, Z^s \\ div\, P^{s\perp} = 0 \\ \text{with } P^{s\perp} n = 0 \text{ on boundary of the body } B \end{array}\right\} \tag{4}$$

It is clear from (3) and (4) that when $\Pi^s = 0$ then $P^{s\perp} = 0$. Thus, referring back to our physical picture of a single terminating elastic distortion discontinuity, the expression

$$G^e = P^{s\perp} + grad\, U^e + grad\, Z^s \tag{5}$$

implies that a smooth representation of elastic distortion discontinuities across surfaces that do not terminate or have discontinuities in their tangent plane orientation field is embodied in the *compatible* part of G^e, characterized by $grad\, U^e + grad\, Z^s$. Thus, the compatible part of G^e characterizes smoothed phase/grain boundaries without kinks or corners. $P^{s\perp}$, on the other hand, characterizes phase/grain boundaries with kinks or corners. Also, recognizing $curl\, U^e = -grad\, U^e : X$ as representative of *slip* dislocations arising from slip discontinuities, the expression

$$\alpha^s = -G^e : X = -P^{s\perp} : X - grad\, U^e : X \\ -grad\, Z^s : X$$

implies that $-grad\, Z^s : X$ is representative of pure *transformation* dislocations, while $-P^{s\perp} : X$ is representative of *g-disclination-induced transformation* dislocations. Thus, it becomes clear why an infinite grain boundary/incoherent phase boundary can often be represented by slip dislocations. In addition, (5) also makes clear what parts of the structure of a general grain/phase boundary cannot be represented by slip dislocations.

We now assume that the dislocation and the g-disclination densities representing line-like objects have velocity fields, V^α and V^π, associated with them. In situations when these fields represent individual defects through non-singular localization in core cylinders, the corresponding local value of the velocity field represents the velocity of the movement of the core cylinders. Furthermore, because these defect fields are line densities, conservation laws for their evolution are an immediate consequence (Acharya 2011). The local form of the conservation law for Π^s is given by

$$\dot{\Pi}^s = -curl\,\left(\Pi^s \times V^\pi\right). \tag{6}$$

As for the evolution of the line density field $curl\, U^e$, we note from (5) that

$$\dot{\alpha}^s + \dot{P}^s : X = curl\, \dot{U}^e \tag{7}$$

which we set equal to $-curl\, (\alpha^s \times V^\alpha)$ from the kinematics of flux of lines moving into an area patch through its bounding curve[3] to obtain

$$\dot{\alpha}^s = -curl\,\left(\alpha^s \times V^\alpha\right) - \dot{P}^s : X, \tag{8}$$

which implies

$$div\, \dot{\alpha}^s = \dot{\Pi}^s : I.$$

Physically, this means that in contrast to the pure dislocation case, dislocation lines can terminate within the body on phase/grain boundaries, but the locations at which they do so necessarily involve a variation in the strength of the elastic distortion discontinuity along the phase/grain boundary.

The statements (3) and (6) imply

$$\dot{P}^s = -\Pi^s \times V^\pi + grad\, K^s, \tag{9}$$

where K^s is a strain rate (2nd-order tensor) and $grad\, K^s$ is a compatible contribution to the elastic 2-distortion rate associated with the transverse motion of (physically identifiable) phase/grain boundaries. K^s requires specification based on geometric and constitutive considerations.

We now introduce the definition

$$U^p := grad\, u - U^e. \tag{10}$$

Then, (7), (8), (10) imply

$$\dot{U}^p = \alpha^s \times V^\alpha, \tag{11}$$

where, unlike the 2-distortion case, we do not include the free gradient (representing an inelastic strain rate) that arises in deducing (11) due to the assumption that slip-boundaries are often not physically identifiable. Even when they are, as in the case of stacking faults, their evolution typically consists of expansion or shrinkage in the slip plane governed by the motion of their bounding dislocation curve(s), so that no generation of independent inelastic strain rate may be assigned to this evolution.

Summarizing, one minimal representation of the field equations of the model consist of Eqs. (1), (2), (3), (9), (11) along with

$$div\, T = 0 \tag{12}$$

$$div\, \Lambda - X : T = 0 \tag{13}$$

representing balance of linear momentum and angular momentum, ignoring body forces and couples and inertia terms (without loss of essential generality for the present purpose). The last two equations serve to determine the displacement field, when appended with constitutive equations for the stress and couple stress tensors. In addition, constitutive equations are also required for the dislocation and g-disclination velocities as well as the potential K^s for the compatible part of the elastic 2-distortion rate, G^e.

5 Constitutive guidance from thermodynamics

Following Mindlin and Tiersten (1962), we consider the following statement to characterize the work of external agents on the body:

[3] An equally plausible expression would be to set $-curl\, (curl\, U^e \times V^\alpha)$, but the one we choose appears to have a definite advantage for the modeling of disclination and dislocation free phase transformations.

$$\int_{\partial\mathcal{B}} (Tn) \cdot v\, dv + \int_{\partial\mathcal{B}} (\Lambda n) \cdot \omega\, dv,$$

where $\mathcal{B}$ is the body, $\partial\mathcal{B}$ its boundary and n the outward unit normal to the boundary. Denoting by ψ the free energy per unit volume field on the body, we define the (mechanical) dissipation to be

$$\begin{aligned} \mathsf{D} &= \int_{\partial\mathcal{B}} (Tn) \cdot v\, dv + \int_{\partial\mathcal{B}} (\Lambda n) \cdot \omega\, dv - \int_{\mathcal{B}} \dot{\psi}\, dv \\ &= \int_{\mathcal{B}} T_{sym} : D\, dv - \frac{1}{2} \int_{\mathcal{B}} \Lambda : grad\,(X : \Omega)\, dv \\ &\quad - \int_{\mathcal{B}} \dot{\psi}\, dv. \end{aligned} \tag{14}$$

We now consider three possible constitutive dependencies for ψ and show that even though 'higher-order' kinematic objects are involved in the specific free energy, in two of these cases no couple stress is required, at least based on thermodynamic considerations. We also derive the driving forces for the dissipative mechanisms of dislocation motion (characterized by V^{α}), g-disclination motion (characterized by V^{π}) and phase/grain boundary motion (characterized by K^s, up to further future geometric refinement to go from a strain-rate to a phase-front velocity).

By substituting these constitutive hypotheses into (14) we identify the dissipative power-conjugates to the defect motion variables as driving forces. We also identify constitutive equations for the stress and couple stress tensors by requiring that $\mathsf{D} = 0$ for all possible motions of the body in the absence of the dissipative mechanisms.

5.1 $\psi = \hat{\psi}\,(U^e, P^s, \Pi^s)$ implies no couple stress

The considered constitutive class is one that contains elastic 2-distortion contributions in the specific free energy arising only from disclination and phase/grain boundary-induced fields, but none from the gradient of the elastic 1-distortion. We require that

$\hat{\psi}\left(U^e, P^s, \Pi^s\right) = \hat{\psi}\left(U^e + W, P^s, \Pi^s\right)\ \forall\ W$ skew,

which implies that $\partial_{U^e}\psi$ has to be symmetric (and more). Also,

$$\begin{aligned} \dot{\psi} &= \left(\partial_{U^e}\hat{\psi}\right)_{sym} : grad\, v - \left(\partial_{U^e}\hat{\psi}\right)_{sym} \\ &\quad : \dot{U}^p + \partial_{P^s}\hat{\psi} : \dot{P}^s + \partial_{\Pi^s}\hat{\psi} : \dot{\Pi}^s \end{aligned}$$

which, along with (6, 9, 11) and (14) implies,

$$\begin{aligned} \mathsf{D} &= \int_{\mathcal{B}} \left[T_{sym} - \left(\partial_{U^e}\hat{\psi}\right)_{sym} \right] : D\, dv \\ &\quad - \frac{1}{2} \int_{\mathcal{B}} \Lambda : grad\,(X : \Omega)\, dv \\ &\quad + \int_{\mathcal{B}} \left[X \left(\partial_{U^e}\hat{\psi}\right)_{sym} \alpha^s \right] \cdot V^{\alpha}\, dv \\ &\quad + \int_{\mathcal{B}} \left[X \left\{ \left(\partial_{P^s}\hat{\psi} + curl\, \partial_{\Pi^s}\hat{\psi}\right)_{ijl} \right.\right. \\ &\quad \left.\left. \Pi^s_{ijk} e_l \otimes e_k \right\} \right] \cdot V^{\pi}\, dv \\ &\quad + \int_{\mathcal{B}} div \left(\partial_{P^s}\hat{\psi}\right) : K^s\, dv \\ &\quad - \int_{\partial\mathcal{B}} \left(K^s : \partial_{P^s}\hat{\psi}\right) n\, da + \int_{\partial\mathcal{B}} \left(\Pi^s \times V^{\pi}\right) \\ &\quad \vdots \left(\partial_{\Pi^s}\hat{\psi} \times n\right) da. \end{aligned} \tag{15}$$

Thus we conclude that

$$T = \left(\partial_{U^e}\hat{\psi}\right)_{sym} \tag{16}$$

$$\Lambda = 0 \tag{17}$$

$$X\left(T^T \alpha^s\right) \hookrightarrow V^{\alpha} \text{ in bulk} \tag{18}$$

$$X\left\{ \left(\partial_{P^s}\hat{\psi} + curl\, \partial_{\Pi^s}\hat{\psi}\right)_{ijl} \Pi^s_{ijk} e_l \otimes e_k \right\} \hookrightarrow V^{\pi} \tag{19}$$

$$K^s \hookrightarrow div\left(\partial_{P^s}\hat{\psi}\right) \text{ in bulk,} \tag{20}$$

form a consistent set of constitutive guidelines that satisfies balance of angular momentum identically. Here $\hookrightarrow$ is used as shorthand for "is the driving force for." In addition, the boundary terms in (15) may be interpreted as providing driving forces for K^s and V^{π} on the boundary of the body.

The driving 'force' in (18) is the analog of the Peach–Koehler force of classical dislocation theory; it has physical units of $force/volume$, to be interpreted as the force acting on unit volume of the dislocation core cylinders. The driving force in (19) is the corresponding thermodynamic force on unit volume along the length of a g-disclination cylinder; this is a new construct whose analog for the pure disclination case arises in Fressengeas et al. (2011). Finally, (20) shows the driving force for the inelastic strain-gradient rate produced due to the motion of phase/grain boundaries.

5.2 $\psi = \hat{\psi}(U^e, P^s, \Pi^s, curl\, U^e)$ implies no couple stress

Define $\tilde{\alpha} = curl\, U^e$. Then, an additional term to the dissipation (15) is

$$-\int_{\mathcal{B}} \partial_{\tilde{\alpha}} \hat{\psi} : \dot{\tilde{\alpha}}\, dv = \int_{\mathcal{B}} \partial_{\tilde{\alpha}} \hat{\psi} : curl\left(\alpha^s \times V^\alpha\right) dv.$$

The results (16), (17), (19) remain unchanged and (18) changes to

$$X\left(T + curl\, \partial_{\tilde{\alpha}} \hat{\psi}\right)^T \alpha^s \hookrightarrow V^\alpha \text{ in bulk,}$$

with an additional boundary term in the dissipation (15) given by

$$-\int_{\partial\mathcal{B}} \partial_{\tilde{\alpha}} \hat{\psi} : \left[\left(\alpha^s \times V^\alpha\right) \times n\right] da.$$

It is interesting to note that the presence of a special combination of $grad\, U^e$, i.e., the part only related to slip dislocation content, allows a theory without couple stress and a symmetric stress tensor on thermodynamic grounds. A special case of the constitutive class considered here was treated, with similar results, in Acharya (2010).

5.3 $\psi = \hat{\psi}(U^e, P^s, \Pi^s, grad\, U^e)$ implies model with couple stress

We use the notation $grad\, U^e =: \nabla U^e$ and observe that a functional dependence of the specific free energy on the whole elastic 2-distortion, i.e., $\psi = \tilde{\psi}(U^e, G^e, \Pi^s)$ is a special case of the constitutive class being considered for which

$$\partial_{\nabla U^e} \hat{\psi}\left(U^e, P^s, \Pi^s, \nabla U^e\right) = \partial_{G^e} \tilde{\psi}\left(U^e, G^e, \Pi^s\right).$$

In this case, the additional terms in the dissipation (15) are

$$-\int_{\mathcal{B}} \partial_{\nabla U^e} \hat{\psi} \,\vdots\, gradgrad\, v\, dv + \int_{\mathcal{B}} \partial_{\nabla U^e} \hat{\psi} \,\vdots\, grad\left(\alpha^s \times V^\alpha\right) dv.$$

Noting that

$$\Lambda : grad\,(X\Omega) = -e_{mni} \Lambda_{ij} v_{m,nj},$$

(17) is replaced by the relation

$$e_{mni} \Lambda_{ij} + e_{mji} \Lambda_{in} = -\left[\left(\partial_{\nabla U^e} \hat{\psi}\right)_{mnj} + \left(\partial_{\nabla U^e} \hat{\psi}\right)_{mjn}\right],$$

(18) is replaced by

$$X\left(T - div\, \partial_{\nabla U^e} \hat{\psi}\right)^T \alpha^s \hookrightarrow V^\alpha \text{ in bulk,}$$

with an additional boundary term in the dissipation (15) given by

$$\int_{\partial\mathcal{B}} \left[\left(\alpha^s \times V^\alpha\right) : \partial_{\nabla U^e} \hat{\psi}\right] n\, da.$$

Thermodynamical considerations leave the following parts of the stress and couple stress tensors,

$$T_{ij} - T_{ji}$$
$$e_{mni} \Lambda_{ij} - e_{mji} \Lambda_{in},$$

constitutively unspecified. Balance of linear momentum (12) and angular momentum (13) along with boundary conditions serve as two constraints for their determination.

6 Concluding remarks

A 'small-deformation' theory of coupled plasticity and phase transformation accounting for the dynamics of generalized defects, i.e., dislocations and g-disclinations, has been proposed. When the elastic strain-gradient is compatible, the present theory reduces to that of Fressengeas et al. (2011), which involves only dislocations and pure disclinations.

A fundamental premise of our work is that it is not necessary, neither efficient, to resolve atomic level vibrations to model defect dynamics leading to inelasticity at meso- and macroscopic scales. Our PDE-based modeling paradigm is to account for the defects in atomic configurations in (material-specific) nonlinear elastic media, focusing on the dynamics of distributions of defects in such configurations, rather than the dynamics of the atoms themselves. This requires minimal, but essential, input on the mobility of single defects and nonlinear (and non-monotone) elastic behavior from MD/Quantum mechanics. Such transfer of information has the potential for computational efficiency because one does not have to resolve atomic vibrations anymore. Our general philosophy is motivated by the great success of the Peierls model in elucidating basic dislocation physics, and is a generalization of that approach to deal with unrestricted geometric nonlinearity, full-fledged non-monotone bulk elasticity and inertia. Moreover, the most noteworthy merit of such a PDE-based approach is that it suggests rational and natural ways for developing coarse-grained meso/macroscale models of averaged defect densities interacting with stress and leading to inelastic flow and failure, a subject of great current interest.

Acknowledgments CF gratefully acknowledges funding from CMU and UPV-M for his visits to CMU. AA acknowledges the hospitality and support of the LEM3, UPV-M, and the T4-Division of the Los Alamos National Laboratory.

References

Acharya A (2010) New inroads in an old subject: plasticity, from around the atomic to the macroscopic scale. J Mech Phys Solids 58(5):766–778

Acharya A (2011) Microcanonical entropy and mesoscale dislocation mechanics and plasticity. J Elast 104:23–44

Abeyaratne R, Knowles JK (1990) On the driving traction acting on a surface of strain discontinuity in a continuum. J Mech Phys Solids 38(3):345–360

Acharya A, Matthies K, Zimmer J (2010) Travelling wave solutions for a quasilinear model of field dislocation mechanics. J Mech Phys Solids 58:2043–2053

Acharya A, Tartar L (2011) On an equation from the theory of field dislocation mechanics. Bollettino dell'Unione Matematica Italiana IV(9): 409–444

de Wit R (1970) Linear theory of static disclinations. Fundam Aspects Dislocat Theory 1:651–673

Fosdick R (2011) Observations concerning virtual power. Math Mech Solids 16(6):573–585

Fressengeas C, Taupin V, Capolungo L (2011) An elasto-plastic theory of dislocation and disclination fields. Int J Solids Struct 48:3499–3509

Gurtin ME (2000) Configurational forces as basic concepts of continuum physics, vol 137. Springer, Berlin

Hirth JP, Pond RC (2011) Compatibility and accommodation in displacive phase transformations. Prog Mater Science 586–636. doi:10.1016/j.pmatsci.2011.01.003

Kleinert H (2008) Multivalued fields in condensed matter, electromagnetism, and gravitation. World Scientific, Singapore

Levitas VI, Levin VA, Zingerman KM, Freiman EI (2009) Displacive phase transitions at large strains: phase-field theory and simulations. Phys Rev Lett 103(2):25702

Mindlin RD, Tiersten HF (1962) Effects of couple-stresses in linear elasticity. Arch Ration Mech Anal 11(1):415–448

Roy A, Acharya A (2005) Finite element approximation of field dislocation mechanics. J Mech Phys Solids 53(1):143–170

Simha NK, Bhattacharya K (1998) Kinetics of phase boundaries with edges and junctions in a three-dimensional multi-phase body. J Mech Phys Solids 46:2323–2359

Int J Fract (2012) 174:95–102
DOI 10.1007/s10704-012-9685-3

ORIGINAL PAPER

Continuum strain-gradient elasticity from discrete valence force field model for diamond-like crystals

A. Danescu · G. Grenet

Received: 17 August 2011 / Accepted: 1 February 2012 / Published online: 29 February 2012

Abstract In this paper using polynomial approximation of the dispersion relations near the center of the first Brillouin zone, we construct a continuum model with internal length scale starting from the discrete valence force field model. As a consequence, and in agreement with a recent result by Maranganti and Sharma, we obtain an explicit estimate of the internal lengths which is $\simeq 0.1$ Å.

Keywords Dispersive corrections of elasticity · Size effects in elasticity · Strain gradient theory

1 Introduction

Recent progress in nanotechnology evidenced various size-effects in both metals and semi-conductors. An interesting overview of size-effects is presented in Park et al. (2009), which concludes that *a substantial gap exists between computational predictions and experimental measuring of both metallic and semi-conductor nano-wires*. It is well-known that the classical theory of elasticity is unable to predict size-effects since it does not incorporate an internal length. Moreover, at very small length scales (at least one characteristic length at nano-scale) the continuum theory is questionable and powerful numerical methods like density functional theory and molecular dynamics provide a valuable tool for investigation of material properties.

The aim of this short paper is to provide a bridge between the discrete model for group IV elements in the diamond structure (C, Si, Ge) and a continuum elasticity theory that includes an internal length. Such a theory can be regarded as a dispersive correction of classical elasticity that involves the strain gradient. From a phenomenological point of view the general framework of second-order elasticity was studied extensively (Mindlin 1965, 1964; Toupin 1962) essentially in the isotropic case. Accounting for the first (or higher) gradient of the strain in the free-energy density introduces additional material parameters. It is known that in the cubic case the coupling between the strain and its gradient introduces one additional material parameter and there are six material parameters in the quadratic part associated with the gradient of the strain. The significant number of additional material parameters and their role were important drawbacks in the development of the strain gradient theory.

From a different perspective progress on molecular dynamics (Stillinger and Weber 1985; Daw and Baskes 1984) improved significantly the discrete modeling of interaction forces and recently (Maranganti and Sharma 2007) combined DFT and MD numerical results have provided a path for numerical estimates of higher-order elastic constants. The originality of the

A. Danescu (✉) · G. Grenet
INL, UMR CNRS 5270, Ecole Centrale de Lyon, Ecully, France
e-mail: danescu@ec-lyon.fr

G. Grenet
e-mail: genevieve.grenet@ec-lyon.fr

present paper consists in the analytical investigation of the relation between the discrete models and continuum models. In other words we obtain analytical expressions of higher-order elasticities from the continuum theory in terms of interaction stiffnesses from the discrete theory and lattice parameter. The method follows the line initiated by Krumhansl (1967), DiVincenzo (1986) and uses lattice dynamics and successive approximations of the acoustic branches of the dispersion relation.

The paper is organized as follows: the second section presents the discrete model and its pertinence with respect to experimental dispersion data. The third section presents lattice dynamics in general and the separation of the acoustic and optical branches. We specialize the results to the first-approximation which leads to classical elasticity theory and to the fourth-order approximation which is the theory including the gradient of strain. In the fourth section we compute the numerical values of the higher-order elasticities and we show that the obtained values are in good agreement with results from Maranganti and Sharma (2007) and estimations in Mindlin (1965).

2 The valence force field and the macroscopic elasticity

The elasticities of group IV elements with diamond-like structure is remarkably well described by the valence force field (Keating 1966). This model was proposed by Keating in an attempt to correct a previous result of Born (1914) obtained using the long wave-length method. The elastic energy accounts for a *pair interactions potential* expressed as

$$w_i^j = \frac{2A}{3a^2}\left((\boldsymbol{x}^j - \boldsymbol{x}^i)\cdot(\boldsymbol{x}^j - \boldsymbol{x}^i) - \frac{3a^2}{16}\right)^2 \tag{1}$$

and a *three-body potential* (angular interaction) in the form

$$w_i^{jk} = \frac{8B}{3a^2}\left((\boldsymbol{x}^j - \boldsymbol{x}^i)\cdot(\boldsymbol{x}^k - \boldsymbol{x}^i) + \frac{a^2}{16}\right)^2. \tag{2}$$

In (1) and (2) a denotes the lattice parameter, $A > 0$ and $B > 0$ are stiffnesses of pair and three-body interactions and $\boldsymbol{x}^i$ denotes the (actual) position of the atom i. The total elastic energy is

$$W = \sum_i \left[\sum_{j\in NN(i)} w_i^j + \sum_{(j,k)\in NN(i)} w_i^{jk}\right]. \tag{3}$$

Table 1 Lattice parameters a, (experimental data from Ioffe Physico-Technical Institute) computed stiffnesses A, B, using experimental data for C_{11} and C_{12} using (4-1) and (4-2); predicted C_{44} using (4-3) and expected (experimental data); the value of the left-hand-side in (5)

	C	Si	Ge
a (Å)	3.567	5.431	5.658
A (N/m)	388.1	145.8	109.5
B (N/m)	63.87	10.38	8.67
C_{11} (GPa)	1079	166	126
C_{12} (GPa)	124	64	44
C_{44}^{th} (GPa)	575.9	79.3	62.2
C_{44}^{exp} (GPa)	578	79.6	67.7
LHS in (5)	1.004	1.002	1.08

where $NN(i)$ denotes the set (indexes) of atoms which are the near-neighbors of the atom i.

Discrete homogenization method (details presented in the Appendix) gives the macroscopic elasticities predicted by the discrete valence force field model

$$C_{11} = \frac{A+12B}{3a}, \quad C_{12} = \frac{A-4B}{3a},$$
$$C_{44} = \frac{16AB}{3a(A+4B)}. \tag{4}$$

The macroscopic Hooke law obtained satisfies the strong ellipticity condition (Chirita et al. 2007). We notice here that the three macroscopic elasticities depend on two microscopic stiffnesses, so that the model also predicts the relation

$$\frac{2C_{44}(C_{11}+C_{12})}{(C_{11}-C_{12})(C_{11}+3C_{12})} = 1. \tag{5}$$

Table 1 presents the macroscopic elasticities for C, Si and Ge the microscopic stiffnesses A and B computed using (4-1) and (4-2) the predicted C_{44} using (4-3) and the value of the ratio in (5). The computed elastic constant C_{44} is in very good agreement with the experimental data, result confirmed by the value of the ratio in the left-hand-side of (5).

However, it is well-known that the elastic constants describe only acoustic branches of the dispersion relation near the center of the first Brillouin zone (BZ). In order to conclude that the valence force field model describes accurately the behavior of the lattice one has to compare the experimental and predicted dispersion relations in the first BZ. The comparaison is presented in Figs. 1 and 2. Except for the discrepancy of one

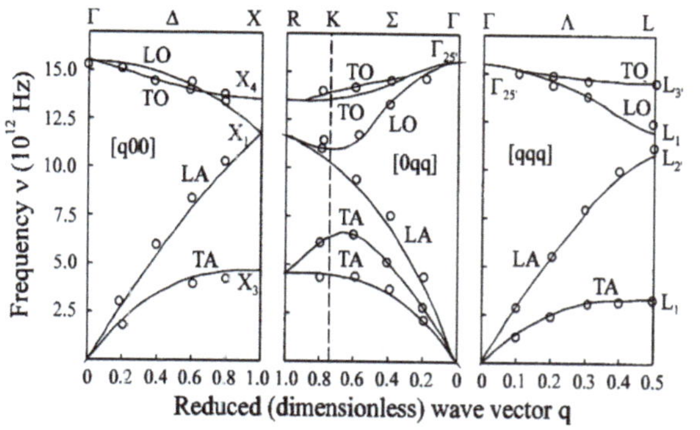

Fig. 1 Experimental data for the phonon spectrum (acoustic optical branches) along the high symmetry *lines* in the first Brillouin zone for bulk silicon (Dolling 1963)

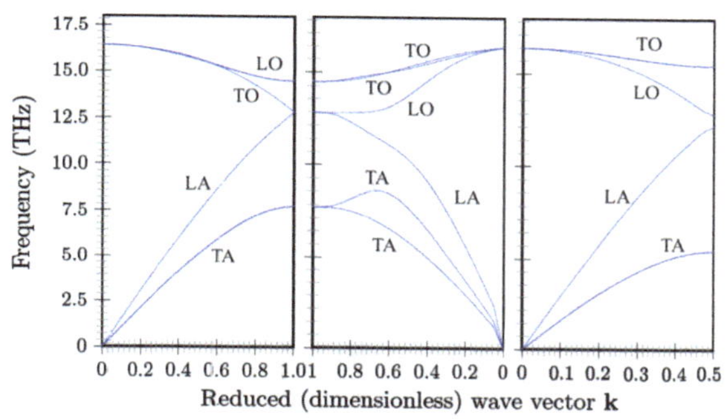

Fig. 2 Computed phonon spectrum (acoustic and optical branches) along the high symmetry *lines* in the first Brillouin zone using the valence force field model

transversal acoustic branch near the boundary of the BZ, near the X (or R) we can conclude that the valence force field agrees with experimental data in the entire first BZ.

The next section presents the dynamics of the lattice and two successive polynomial approximations. The first approximation leads to the classical elasticity while the second one leads to an elastic theory including strain-gradients.

3 Lattice dynamics

We recall that the unit cell of the diamond lattice is non-primitive and contains two identical atoms. Dynamics of the (infinite) lattice is governed by

$$m\ddot{u}^i = -\frac{\partial W}{\partial \boldsymbol{u}^i}. \tag{6}$$

Using the explicit form of the interaction potential we obtain

$$\frac{\partial W}{\partial \boldsymbol{u}^i} = \sum_{j\in NN(i)} \frac{\partial W_i^j}{\partial \boldsymbol{u}^i} + \sum_{\substack{\{j,k\}\subset NN(i)\\ k\neq j}} \frac{\partial W_i^{jk}}{\partial \boldsymbol{u}^i} + \sum_{\substack{j\in NN(i)\\ k\in NN(j)\\ k\neq i}} \frac{\partial W_j^{ik}}{\partial \boldsymbol{u}^i} \tag{7}$$

where $NN(i)$ denotes the nearest-neighbors of atom i. Next step is to look for solutions in the form

$$\boldsymbol{u} = \boldsymbol{U}e^{i(\boldsymbol{k}\cdot\boldsymbol{x}+\omega t)} \tag{8}$$

using $\boldsymbol{U}$ for the amplitude in one of the sublattices and $\boldsymbol{V}$ for the other one. The final result is an homogeneous system of equations for $(\boldsymbol{U}, \boldsymbol{V})$ in the form

$$\left(\begin{bmatrix} \mathbf{A}_{UU}(\boldsymbol{k}) & \mathbf{A}_{UV}(\boldsymbol{k}) \\ \mathbf{A}^{\star}_{UV}(\boldsymbol{k}) & \mathbf{A}_{VV}(\boldsymbol{k}) \end{bmatrix} - \omega^2 m \boldsymbol{I}_{6\times 6}\right)\begin{pmatrix}\boldsymbol{U}\\ \boldsymbol{V}\end{pmatrix} = \begin{pmatrix}\mathbf{0}\\ \mathbf{0}\end{pmatrix} \tag{9}$$

For each $\boldsymbol{k}$ in the first BZ we have 6 solutions $\omega(\boldsymbol{k})$ among which 3 are *acoustic branches* corresponding to the *average* motions in the unit cell and 3 are *optical branches* corresponding to the *relative* motion of two atoms in the unit cell. Invariance of the interaction energy with respect to translation shows that with respect to a suitable basis obtained using *collective variables* $(\boldsymbol{U}+\boldsymbol{V})/\sqrt{2}$ and $(\boldsymbol{U}-\boldsymbol{V})/\sqrt{2}$ we have

$$\boldsymbol{A}(\boldsymbol{k}) = \begin{pmatrix} \boldsymbol{A}_{aa}(\boldsymbol{k}) & \boldsymbol{A}_{ao}(\boldsymbol{k}) \\ \boldsymbol{A}^{\star}_{ao}(\boldsymbol{k}) & \boldsymbol{A}_{oo}(\boldsymbol{k}) \end{pmatrix} = \begin{pmatrix} O(|\boldsymbol{k}|^2) & O(|\boldsymbol{k}|) \\ O(|\boldsymbol{k}|) & O(1) \end{pmatrix}. \tag{10}$$

Using a result of Mielke (2006) the Hooke tensor is then obtained as

$$\hat{\mathbb{C}}[\boldsymbol{v}\otimes\boldsymbol{k}_1]\boldsymbol{k}_2 = \Big[\frac{1}{2}D^2\boldsymbol{A}_{aa}(\mathbf{0})[\boldsymbol{k}_1,\boldsymbol{k}_2] - (D\boldsymbol{A}_{ao}(\mathbf{0})[\boldsymbol{k}_1])\,\boldsymbol{A}_{oo}^{-1}(\mathbf{0})\,\big(D\boldsymbol{A}^{\star}_{ao}(\mathbf{0})[\boldsymbol{k}_2]\big)\Big]\boldsymbol{v} \tag{11}$$

A first consequence of this result is the following: to obtain the classical elastic constants it is sufficient to perform the computations using only the terms of order lower than $O(|\boldsymbol{k}|^3)$ as $\boldsymbol{k}\to\mathbf{0}$. A long but straightforward computation provides the blocks of the dispersion matrix $\boldsymbol{A}(\boldsymbol{k})$ in the form

$$\boldsymbol{A}_{UU} = \frac{4}{3}(A+4B)\boldsymbol{I} + \frac{a^2B}{3}\boldsymbol{k}\otimes\boldsymbol{k} = \boldsymbol{A}_{VV} \tag{12}$$

$$\boldsymbol{A}_{UV} = \frac{A+4B}{3}\left[-4\boldsymbol{I} + \frac{a^2}{8}\boldsymbol{k}\otimes\boldsymbol{k}\right] + \frac{ia(A-4B)}{3}\left[\boldsymbol{S} + \frac{a}{4}\boldsymbol{P}\right] \tag{13}$$

where

$$P=\begin{pmatrix} 0 & k_1k_2 & k_1k_3 \\ k_1k_2 & 0 & k_2k_3 \\ k_1k_3 & k_2k_3 & 0 \end{pmatrix}, \quad S=\begin{pmatrix} 0 & k_3 & k_2 \\ k_3 & 0 & k_1 \\ k_2 & k_1 & 0 \end{pmatrix}. \quad (14)$$

We notice that relation (10) (obtained using collective variables) does not separate the acoustic and optical branches. For classical elasticity this separation is performed by the result in (11). An explicit computation using (12), (13) and (11) leads to the result presented in (4) but since we are interested in higher order approximation we shall follow a different approach.

3.1 Separation of acoustic and optical branches

As we are interested in higher-order approximation of acoustic branches near the center of the first BZ we shall follow a perturbation approach previously developed in DiVincenzo (1986). The main difficulty is the separation of the acoustic and optical branches of the dispersion matrix $\boldsymbol{A}(\boldsymbol{k})$ which is approximation dependent.

For the perturbation method we rewrite (10) as

$$\boldsymbol{A}(\boldsymbol{k}) = \boldsymbol{H} + \boldsymbol{Y}(\boldsymbol{k}) + \boldsymbol{X}(\boldsymbol{k}), \quad (15)$$

where

$$\boldsymbol{H} = \begin{pmatrix} E_a \boldsymbol{I}_{3\times3} & \boldsymbol{0}_{3\times3} \\ \boldsymbol{0}_{3\times3} & E_o \boldsymbol{I}_{3\times3} \end{pmatrix}, \quad (16)$$

$$\boldsymbol{Y}(\boldsymbol{k})=\begin{pmatrix} A_{aa}(\boldsymbol{k})-E_a\boldsymbol{I}_{3\times3} & \boldsymbol{0}_{3\times3} \\ \boldsymbol{0}_{3\times3} & A_{oo}(\boldsymbol{k})-E_o\boldsymbol{I}_{3\times3} \end{pmatrix}, \quad (17)$$

and

$$\boldsymbol{X}(\boldsymbol{k}) = \begin{pmatrix} \boldsymbol{0}_{3\times3} & \boldsymbol{A}_{ao}(\boldsymbol{k}) \\ \boldsymbol{A}^{\star}_{ao}(\boldsymbol{k}) & \boldsymbol{0}_{3\times3} \end{pmatrix} \quad (18)$$

and consider $\boldsymbol{X}(\boldsymbol{k})$ as the small parameter.

In DiVincenzo (1986) it is proved that the splitting between the acoustic and optical branches of successive polynomial approximations $\boldsymbol{A}^{(m)}(\boldsymbol{k})$ of degree $m \le 4$ of the exact dispersion matrix $\boldsymbol{A}(\boldsymbol{k})$ is obtained as

$$\boldsymbol{A}^{(m)}(\boldsymbol{k}) = \sum_{s=0}^{m} \boldsymbol{K}^{(s)} \quad (19)$$

where

$$\begin{aligned} \boldsymbol{K}^{(0)} &= \boldsymbol{H}, \\ \boldsymbol{K}^{(1)} &= \boldsymbol{Y}, \\ \boldsymbol{K}^{(2)} &= \frac{1}{2}[\boldsymbol{X}^H, \boldsymbol{X}], \\ \boldsymbol{K}^{(3)} &= \frac{1}{2}[\boldsymbol{X}, [\boldsymbol{Y}, \boldsymbol{X}^H]^H], \\ \boldsymbol{K}^{(4)} &= \frac{1}{2}\Big[\boldsymbol{X}, \Big\{[[\boldsymbol{Y}, \boldsymbol{X}^H]^H, \boldsymbol{Y}]^H \\ &\quad + \frac{2}{3}[\boldsymbol{K}^{(2)}, \boldsymbol{X}^H]^H\Big\}\Big] - \frac{1}{12}[[\boldsymbol{K}^{(2)}, \boldsymbol{X}^H], \boldsymbol{X}^H], \end{aligned} \quad (20)$$

the commutator $[\boldsymbol{X}, \boldsymbol{Y}]$ denotes $\boldsymbol{XY} - \boldsymbol{YX}$ and $\boldsymbol{X}^H$ is the (unique) solution $\boldsymbol{A}$ of the equation $[\boldsymbol{H}, \boldsymbol{A}] = \boldsymbol{X}$. It follows that, as $\boldsymbol{k} \to \boldsymbol{0}$ we have

$$\boldsymbol{A}(\boldsymbol{k}) = \underbrace{\begin{pmatrix} \boldsymbol{D}^{(m)}_{aa}(\boldsymbol{k}) & \boldsymbol{0}_{3\times3} \\ \boldsymbol{0}_{3\times3} & \cdots \end{pmatrix}}_{\boldsymbol{A}^{(m)}(\boldsymbol{k})} + O(|\boldsymbol{k}|^{m+1}) \quad (21)$$

where $\boldsymbol{D}^{(m)}_{aa}$ is a polynomial of degree $\le m$. Identification with classical dispersion relations from the continuum setting will provide the macroscopic approximation. Obviously, cubic symmetry is inherited from the discrete setting.

As previously noted, in order to obtain $\boldsymbol{K}^{(m)}$ it is sufficient to perform computations with truncated Taylor series at $O(|\boldsymbol{k}|^{m+1})$ as $\boldsymbol{k} \to \boldsymbol{0}$. Taking into account (12) and (13) we obtain

$$\boldsymbol{H} = \frac{8}{3}(A+4B)\begin{pmatrix} \boldsymbol{0}_{3\times3} & \boldsymbol{0}_{3\times3} \\ \boldsymbol{0}_{3\times3} & \boldsymbol{I}_{3\times3} \end{pmatrix}, \quad (22)$$

or, otherwise stated $E_a = 0$ and $E_o = \frac{8}{3}(A+4B)$.

The next subsections provide two successive approximations. The first one is the classical elasticity and, as expected, we shall recover the result in (4). The second approximation is the strain-gradient theory which includes several internal lengths.

3.2 First approximation: classical elasticity

A long but straightforward calculation shows that, taking into account (16) and (22) we obtain at $O(|\boldsymbol{k}|^3)$ when $\boldsymbol{k} \to \boldsymbol{0}$

$$\boldsymbol{H} = \begin{pmatrix} \boldsymbol{0}_{3\times3} & \boldsymbol{0}_{3\times3} \\ \boldsymbol{0}_{3\times3} & \cdots \end{pmatrix} \quad (23)$$

and

$$\boldsymbol{D}^{(2)}_{aa} = \begin{pmatrix} d^{(2)}_{11} & d^{(2)}_{12} & \cdots \\ d^{(2)}_{12} & \cdots & \cdots \\ \cdots & \cdots & \cdots \end{pmatrix} \quad (24)$$

where

$$d^{(2)}_{11} = \frac{a^2\left[(A^2+16AB+48B^2)k_1^2+16AB(k_2^2+k_3^2)\right]}{24(A+4B)}$$

$$d^{(2)}_{12} = \frac{a^2\left[(A^2+16AB-16B^2)k_1k_2\right]}{24(A+4B)}$$

The unit cell of the diamond structure has volume $a^3/4$ and contains two atoms we identify

$$d_{11}^{(2)} = \frac{a^3}{8}\left[C_{11}k_1^2 + C_{44}\left(k_2^2+k_3^2\right)\right]$$
$$d_{12}^{(2)} = \frac{a^3}{8}C_{12}k_1k_2$$

which, as expected, gives the discrete-to-classical elasticity relations (4).

3.3 Second approximation: strain-gradient theory

The next approximation shows that at $O(|\boldsymbol{k}|^4)$ as $\boldsymbol{k}\to\boldsymbol{0}$ we have $\boldsymbol{K}^{(3)}=\boldsymbol{0}$. This means that there is *no coupling* between the strain and its gradient in the free-energy, a result that can be obtained also using group representation theory and the fact that the diamond structure is centro-symmetric.

Computation of $\boldsymbol{K}^{(4)}$ provides a fourth-order homogeneous polynomial correction to the acoustic part of the dispersion matrix. Using group representation arguments (DiVincenzo 1986) one can show that the fourth-order correction can be written in the form

$$\boldsymbol{D}^{(4)} = \begin{pmatrix} \alpha(k_1,k_2,k_3) & \beta(k_1,k_2,k_3) & \beta(k_3,k_1,k_2) \\ \beta(k_1,k_2,k_3) & \alpha(k_2,k_3,k_1) & \beta(k_2,k_3,k_1) \\ \beta(k_3,k_1,k_2) & \beta(k_2,k_3,k_1) & \alpha(k_3,k_1,k_2) \end{pmatrix} \tag{25}$$

where

$$\alpha(k_1,k_2,k_3) = \hat{f}_{111111}k_1^4 + \hat{f}_{122122}\left(k_2^4+k_3^4\right) + 6\hat{f}_{122133}k_2^2k_3^2 + 6\hat{f}_{211222}k_1^2\left(k_2^2+k_3^2\right)$$
$$\beta(k_1,k_2,k_3) = k_1k_2\left[4\hat{f}_{112222}\left(k_1^2+k_2^2\right) + 12\hat{f}_{112233}k_3^2\right].$$

This result shows that for centro-symmetric crystals with cubic symmetry the strain-gradient theory introduce six additional elastic parameters. A long but straightforward computation using (20-4) shows that at $O(|\boldsymbol{k}|^5)$ as $\boldsymbol{k}\to\boldsymbol{0}$ we have

$$\hat{f}_{111111} = 0,$$
$$\hat{f}_{122122} = -a^4\frac{(A-4B)^2(7A^2+40AB+112B^2)}{4608(A+4B)^3},$$
$$\hat{f}_{122133} = -a^4\frac{(A-4B)^3(13A+44B)}{13824(A+4B)^3},$$
$$\hat{f}_{211222} = -a^4\frac{(A-4B)^2(13A^2+88AB+208B^2)}{27648(A+4B)^3},$$
$$\hat{f}_{122222} = -a^4\frac{(A-4B)^2(5A^2+32AB+80B^2)}{9216(A+4B)^3},$$
$$\hat{f}_{112233} = \hat{f}_{122133}. \tag{26}$$

On the other-side the phenomenological theory including strain gradients starts with the expression of the free-energy density in the form

$$w = \frac{1}{2}C_{ijkl}u_{i,j}u_{k,l} + F^1_{ijklmn}u_{i,j}u_{k,lmn} + F^2_{ijklmn}u_{i,jk}u_{l,mn} \tag{27}$$

and leads to a dispersion relation expressed using the dispersion matrix

$$D_{ij} = C_{ikjl}k_kk_l + f_{ikljmn}k_kk_lk_mk_n \tag{28}$$

where

$$f_{ijklmn} = \mathrm{sym}_{(i,k)}\mathrm{sym}_{(j,l,m,n)}(F^1-F^2)_{ijklmn}. \tag{29}$$

Relation (28) needs several remarks.

- Relation (26-1) shows that there is no dispersion for longitudinal waves along the high symmetry axes.
- It is known that there is a one-to-one correspondence between the dispersion matrix and the Hooke law as far as *only* the classical elasticity is accounted for. Relation (29) shows that the one-to-one correspondence is lost if one considers the next polynomial terms in the dispersion matrix (DiVincenzo 1986).
- Previous work in Metrikine and Askes (2002) shows that for the one-dimensional chain with NN interactions the fourth-order approximation near $k=0$ leads to the so-called *sign paradox* which concerns the sign of the k^4 term in the Taylor series and leads to un-physical exponential growth of instabilities at short wavelengths (Metrikine and Askes 2002). This can be avoided in the multi-dimensional case as a consequence of (29) as discussed in Maranganti and Sharma (2007).
- Relations (26) provide higher-order elastic constants in terms of stiffnesses A and B. Along with the lattice parameter a these are the only material parameters in the discrete setting and we have shown in (4) that there is a one-to-one relation between (A,B) and (C_{11},C_{12}). Together these results show that, due to the actual simplicity of the discrete model, the higher-order elasticities depend on classical elastic parameters, a relation which may be regarded as an higher-dimensional analog[1] of (5).

[1] This relation can be regarded as an analog of the well-known Cauchy relations.

Table 2 Numerical results for the second-order elasticities obtained using (26) and numerical values for A and B from Table 1

	C	Si	Ge
f_{111111} (nN)	0	0	0
f_{122122} (nN)	−0.10	−0.35	−0.24
f_{122133} (nN)	−0.01	−0.13	−0.08
f_{211222} (nN)	−0.03	−0.11	−0.08
f_{112222} (nN)	−0.04	−0.13	−0.09
f_{112233} (nN)	−0.01	−0.13	−0.08

Taking into account the volume per atom in the diamond lattice, expressed as functions of C_{11} and C_{12} the second-order elastic constants are

$$\begin{aligned}
\hat{f}_{111111} &= 0,\\
\hat{f}_{122122} &= -a^2\frac{C_{12}^2(3C_{11}^2+4C_{11}C_{12}+5C_{12}^2)}{48(C_{11}+C_{12})^3},\\
\hat{f}_{122133} &= -a^2\frac{C_{12}^3(6C_{11}+7C_{12})}{72(C_{11}+C_{12})^3},\\
\hat{f}_{211222} &= -a^2\frac{C_{12}^2(3C_{11}^2+6C_{11}C_{12}+4C_{12}^2)}{144(C_{11}+C_{12})^3},\\
\hat{f}_{112222} &= -a^2\frac{C_{12}^2(9C_{11}^2+18C_{11}C_{12}+13C_{12}^2)}{384(C_{11}+C_{12})^3},\\
\hat{f}_{112233} &= \hat{f}_{122133}.
\end{aligned}\tag{30}$$

which give also the relation

$$f_{122122} = 2(4f_{112222} - 3f_{211222}). \tag{31}$$

The numerical data in Table 1 and relations (26) (or (30-1)–(30-6)) allow numerical estimates for the second-order elasticities.

4 Numerical results

Table 2 provides numerical results in nN for the second-order elastic constants using numerical data from Table (1) for C, Si and Ge.

The polynomial approximations of the dispersion relations can be illustrated by plotting the approximate dispersion relations along the high symmetry axes in the first BZ.[2] Figure 3 shows the restriction to the first BZ of the polynomial approximation representing the

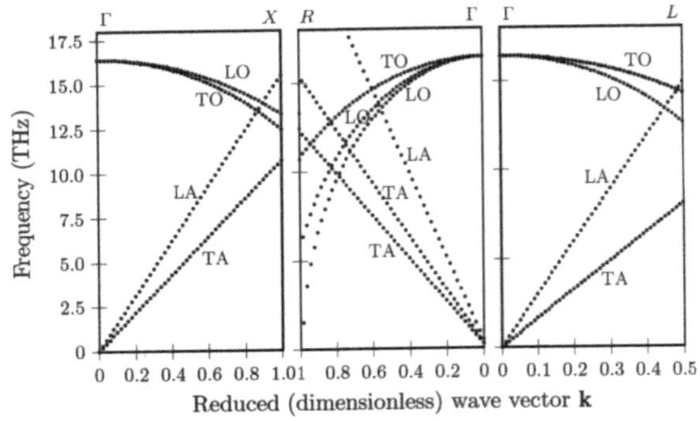

Fig. 3 Restrictions to the high-symmetry *lines* in the first Brillouin zone of the phonon spectrum computed using the first-order approximation of the dispersion matrix. In this approximation the acoustic branches (all starting from origin) are *straight lines*

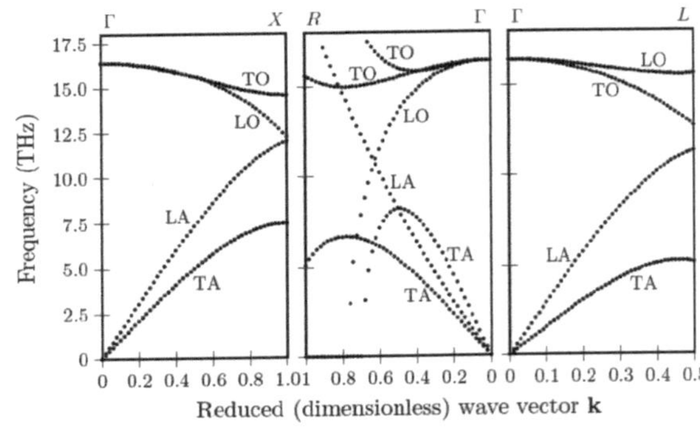

Fig. 4 Restrictions to the high-symmetry *lines* in the first Brillouin zone of the phonon spectrum computed using the second-order approximation of the dispersion matrix. We note that following the behavior of the acoustic branches in the initial model (see Fig. 2), in this approximation, the acoustic branches turn downward. This behavior represents a serious drawback of the approximation, since it leads to loss of positivity of the elastic energy density

classical elasticity for silicon. As expected, the acoustic branches are straight lines and optical branches behave like $\sqrt{\beta^2 - k^2}$. Figure 4 shows the restriction to the first BZ of the next polynomial approximation which, in the real space, corresponds to the theory including the gradient of the strain. In this case, the minus sign in front of the second order elasticities leads to the curving downward of the acoustic branches. This fact leads to the fore-mentioned instabilities at short wavelengths. In Fig. 4 we have only represented the real part of the frequency vs the wavelength k along the high-symmetry axes (and restricted to the range of the exact discrete model).

[2] Obviously, in this case the dispersion matrix will not be periodic, as is the case in the discrete setting.

Using the numerical data from Table 2 and the values of the classical elastic constants we can estime the internal length(s) of the theory which is given by $l \simeq (f_{ijklmn}/C_{ijkl})^{1/2}$.

Since the second-order elastic constants are $\simeq$0.1 nN and the classical elastic coefficients are $\simeq$100 GPa we obtain an internal length $\simeq$0.1 Å. This is one order of magnitude lower than the internal length obtained using combined DFT and MD in Maranganti and Sharma (2007) but in agreement with an early estimate by Mindlin (1965).

5 Conclusions

In this paper we present an analytical derivation of a dispersive correction of elasticity for diamond-like crystals starting from the discrete valence force field model. The pertinence of the starting point is two-folded: first, it describes the macroscopic elastic behavior extremely well and second, the dispersion relations predicted by the model agree with the experimental dispersion relations in the entire first Brillouin zone.

The method we follow uses lattice dynamics and successive polynomial approximations of the acoustic part of the dispersion relation near the center of the Brillouin zone. The symmetry properties of the discrete model are inherited by the successive approximations in the Fourier space and lead, as expected, to macroscopic models of classical elasticity and to one of its dispersive corrections using the first gradient of strain. The obtained strain gradient theory shares a well-known drawback of the Taylor expansion near the Γ-point which leads in the physical space to instabilities at short wavelengths. The main difficulty of the method is the approximation-order dependent separation between the acoustic and optical branches near the center of the Brillouin zone.

The obtained analytical results show that there is no dispersion for the longitudinal waves propagating in the high-symmetry axes. The characteristic length induced by the presence of the gradient of strain leads to a numerical estimate of the internal length of 0.1 Å, in agreement with a previous numerical result obtained in Maranganti and Sharma (2007) and Mindlin (1965).

Possible extensions of our results concern other types of multi-dimensional approximations of the dispersion relations, among which we cite: rational approximations, Padé approximation, multi-point Taylor methods, etc. A different line for introducing an internal length as a correction of classical elasticity is provided by the one-dimensional discrete model with hyper-pre-stress presented in Charlotte and Truskinovsky (2002). An interesting question is then the relation between the present theory and a model with hyper-pre-stress for diamond like crystals, an issue which will be addressed in a future work.

Acknowledgments This work was partially supported by the programme SPAME (ANR06–0274) founded by the French National Research Agency.

Appendix

We present here the computation leading to 4. It is well-known that the unit cell of the diamond lattice contains two (identical) atoms. First step is to compute the quadratic approximation in the displacement for the valence force field energy. We have

$$\boldsymbol{x}^i = \boldsymbol{X}^i + \boldsymbol{u}^i, \tag{32}$$

thus $\boldsymbol{x}^j - \boldsymbol{x}^i = a\frac{\sqrt{3}}{4}\boldsymbol{n}^{ij} + (\boldsymbol{u}^j - \boldsymbol{u}^i)$. We obtain the *quadratic approximations* of interactions in the valence force field as

$$W_i^j \simeq \frac{A}{2}[(\boldsymbol{u}^j - \boldsymbol{u}^i)\cdot \boldsymbol{n}^{ij}]^2 \tag{33}$$

and

$$W_i^{jk} \simeq \frac{B}{2}[(\boldsymbol{u}^j - \boldsymbol{u}^i)\cdot \boldsymbol{n}^{ki} + (\boldsymbol{u}^k - \boldsymbol{u}^i)\cdot \boldsymbol{n}^{ji}]^2. \tag{34}$$

Since the elastic energy is invariant with respect to translation, under fixed macroscopic deformation $\boldsymbol{E}$, the relative displacement between two NN atoms will be given by

$$\boldsymbol{u}^j - \boldsymbol{u}^i = \boldsymbol{E}(\boldsymbol{X}^j - \boldsymbol{X}^j) + \boldsymbol{U}$$

where $\boldsymbol{U}$ is *the fluctuation*.

Elastic energy per unit cell volume is the sum between the contributions of the four terms accounting for the pair interactions and the twelve terms accounting for three-body interactions, thus we obtain

$$\begin{aligned} W = \frac{1}{24vol(Y)}\Big[& a^2(A+12B)(E_{11}^2 + E_{22}^2 + E_{33}^2) \\ & +2a^2(A-4B)(E_{11}E_{22} + E_{22}E_{33} + E_{11}E_{33}) \\ & +4a^2(A+4B)(E_{12}^2 + E_{13}^2 + E_{23}^2) \\ & +16a(A-4B)(E_{23}U_x + E_{13}U_y + E_{12}U_z) \\ & +16(A+4B)(U_x^2 + U_y^2 + U_z^2)\Big]. \end{aligned} \tag{35}$$

Minimization with respect to the relative displacement between the two-lattices $\boldsymbol{U}$ (which represents equilibrium of one of the atoms in the unit cell) gives

$$\boldsymbol{U} = -\frac{a}{2}\frac{A-4B}{A+4B}\begin{pmatrix} E_{23} \\ E_{13} \\ E_{12} \end{pmatrix}. \tag{36}$$

Substitution of this result in (35) gives a quadratic form (in $\boldsymbol{E}$)

$$\begin{aligned} W_{macro.}(\boldsymbol{E}) &= \frac{1}{vol(Y)} \min_{\boldsymbol{U}} W_{micro}(\boldsymbol{E},\boldsymbol{U}) \\ &= \frac{1}{6a(A+4B)} \left[A^2(E_{11}+E_{22}+E_{33})^2 \right. \\ &\quad +16AB\left(E_{11}^2+E_{22}^2+E_{33}^2\right) \\ &\quad +4\left(E_{12}^2+E_{13}^2++E_{23}^2\right) \\ &\quad +16B^2\left(3\left(E_{11}^2+E_{22}^2+E_{33}^2\right)\right. \\ &\quad \left.\left. -2(E_{11}E_{22}+E_{11}E_{33}+E_{22}E_{33})\right)\right], \end{aligned} \tag{37}$$

whose coefficients provide the macroscopic elastic constant given in (4) and completes the proof.

References

Born M (1914) Zur Raumgittertheorie des Diamanten. Annalen der Physik 44:605

Charlotte M, Truskinovsky L (2002) Linear elastic chain with a hyper-pre-stress. J Mech Phys Solids 50(2):217

Chirita S, Danescu A, Ciarletta M (2007) On the strong ellipticity of the anisotropic linearly elastic materials. J Elast 87(1):1

Daw M, Baskes M (1984) Embedded-atom method: derivation and application to impurities, surfaces, and other defects in metals. Phys Rev B 29(12):6443

DiVincenzo DP (1986) Dispersive corrections to continuum elastic theory in cubic crystals. Phys Rev B 34(8):5450

Dolling G (1963) Inelastic scattering neutrons in solids and liquids (proceeding symposium). Chalk River 2:37

Ioffe Physico-Technical Institute database. http://www.ioffe.ru/SVA/NSM/

Keating PN (1966) Effect of invariance requirements on the elastic strain energy of crystals with application to the diamond structure. Phys Rev 145(2):637

Krumhansl J (1967) Some considerations of the relation between solid state physics and generalized continuum mechanics. In: Kroner E (ed) Mechanics of generalized continua. Proceedings of IUTAM symposium on the generalized Cosserat continuum theory of dislocations with applications, Springer, pp 298–311

Maranganti R, Sharma P (2007) A novel atomistic approach to determine strain-gradient elasticity constants: tabulation and comparison for various metals, semiconductors, silica, polymers and the (ir) relevance for nanotechnologies. J Mech Phys Solids 55(9):1823

Metrikine AV, Askes H (2002) One-dimensional dynamically consistent gradient elasticity models derived from a discrete microstructure. Part 1: generic formulation. Eur J Mech A/Solids 21(4):555

Mielke A (2006) Macroscopic behavior of microscopic oscillations in harmonic lattices via Wigner-Husimi transforms. Arch Ration Mech Anal 181(3):401

Mindlin R (1965) Second gradient of strain and surface tension in linear elasticity. Int J Solids Struct 1:417

Mindlin R (1964) Microstructure in linear elasticity. Arch Ration Mech Anal 16:51

Park HEHS, Cai W, Huang H (2009) Mechanics of crystalline nanowires. MRS Bull 34:178

Stillinger F, Weber TA (1985) Computer simulation of local order in condensed phases of silicon. Phys Rev B 31:5262

Toupin R (1962) Elastic materials with couple stresses. Arch Ration Mech Anal 11:385

MIX
Papier aus verantwortungsvollen Quellen
Paper from responsible sources
FSC® C105338

If you have any concerns about our products,
you can contact us on
ProductSafety@springernature.com

In case Publisher is established outside the EU,
the EU authorized representative is:
Springer Nature Customer Service Center GmbH
Europaplatz 3, 69115 Heidelberg, Germany

Printed by Libri Plureos GmbH
in Hamburg, Germany